# Waste to Value Innovations

Motaung Tshwafo Elias
Editor

# Waste to Value Innovations

## Circular Economy for the Energy Sector

Springer

*Editor*
Motaung Tshwafo Elias
Sefako Makgatho Health Sciences
University
Ga-Rankuwa, South Africa

ISBN 978-981-96-2781-3 ISBN 978-981-96-2782-0 (eBook)
https://doi.org/10.1007/978-981-96-2782-0



This Springer imprint is published by the registered company Springer Nature Singapore Pte Ltd.
The registered company address is: 152 Beach Road, #21-01/04 Gateway East, Singapore 189721, Singapore

# Preface

Professor Motaung Tshwafo is a distinguished academic and researcher currently serving as Senior Professor in the Department of Chemistry at Sefako Makgatho Health Sciences University, with a profound commitment to translating scientific and indigenous knowledge into impactful innovations that influence economic growth. In fact, his research philosophy exemplifies the integration of science and societal development.

His expertise lies in organic polymer chemistry, physical chemistry, and nanocomposites. The intersection of science and entrepreneurship is his passion. Among his notable innovations is the development of a bio-brick, a prototype that has garnered attention across the African continent for its potential impact on sustainable construction. A staunch advocate for innovation that empowers small enterprises, Prof. Motaung ensures that his projects align with South Africa's National Development Plan and address critical national priorities.

In addition to his research and innovation contributions, he coordinates a major national collaboration symposium established in 2014. This initiative champions the role of African scholars in advancing education, promoting equitable decolonization, and fostering access to the mainstream economy.

Professor Motaung's work exemplifies the synergy between academic excellence, societal impact, and economic empowerment, making him a key figure in both the academic and innovation landscapes of Africa.

## Industrial Waste Management: A Path to Sustainable Practices

Industrial waste management is critical for promoting sustainable industrial practices, protecting ecosystems, and supporting the global transition to a circular economy. Most industries distributing energy, whether in the form of electricity or gas, often face significant financial inefficiencies stemming from overpayments to subcontractors managing auxiliary aspects of their main projects. This linear approach to resource management could be re-imagined by integrating all potential assets, including by-products and waste streams, into the energy production cycle. Untapped waste resources hold significant potential that has yet to be fully explored.

Why not direct research, innovation, and development efforts toward converting these exhausted materials into new sources of energy or other valuable outputs? This book serves as a guide for investors, offering insights into lucrative opportunities in renewable energy and waste management. It provides a detailed overview of emerging economic streams and forecasts for Africa and beyond. Investors, both individual and collective, will find value in understanding where to channel their resources for maximum returns. The book's content effectively aims to minimize environmental impact, ensure regulatory compliance, and promote sustainability.

## Key Topics Covered

1. Waste Resource Utilization
   - Management of electronic waste, food waste, and exhausted tire waste.
   - Conversion of solid wastes into various products.
   - Assessment of policy directions impacted by waste management.

2. Energy Sector Analysis
   - Examination of global energy sources such as coal, oil, nuclear energy, renewables, natural gas, hydropower, petroleum, solar PVs, and wind power.
   - Insights into the dominance of coal as a global energy source and opportunities for renewable alternatives.

3. Practical Innovations

    - Highlighting industrial waste as a host of chemicals for small-scale laboratory experiments and commercial energy applications.

## Audience and Benefits

- **Academics, Government, and Innovators**: Recommendations for translating research into practical solutions for renewable energy and waste utilization.
- **Ordinary Citizens and Policymakers**: Information on current policies and regulations aimed at accelerating the adoption of innovative energy technologies.
- **Entrepreneurs and Researchers**: Exploration of economic opportunities and actionable strategies for waste-to-energy projects.

## Response to Global Challenges

The COVID-19 pandemic has underscored the need to prioritize practical projects over basic or benchtop research to tackle pressing global challenges effectively. This book captures that transition, presenting a blueprint for transforming knowledge into economic growth and employment opportunities. It identifies gaps in waste and energy management across South Africa and other selected developing countries, offering actionable recommendations for municipalities and relevant stakeholders worldwide to refine their strategies.

## Driving a Knowledge-Based Economy

By highlighting these opportunities, this book inspires a global audience—investors, researchers, academics, and policymakers alike—to embrace a future driven by renewable energy and sustainable waste management. It stands as a beacon for transitioning toward a knowledge-based economy, where innovation leads to tangible benefits for communities and industries.

## Key Takeaways

- **Economic Growth**: Practical recommendations for leveraging waste resources to create new economic streams.

- **Job Creation**: Insights into how waste management innovations can generate employment opportunities.
- **Quality of Life**: Strategies for improving industrial practices to benefit communities and ecosystems.

This book provides direction, ideas, and actionable insights to foster a shift where innovation is the core driver. The proposed transition will lead to tangible benefits such as job creation, improved industries, and enhanced quality of life for communities, leveraging advancements in science, technology, and research.

Ga-Rankuwa, South Africa Motaung Tshwafo Elias

# Contents

# Recycling Electronic Waste (e-Waste)

**Tshwafo E. Motaung**

**Abstract** The electronic industry is one of the fastest growing industries in the world given the fourth industrial revolution. The products of electronic industry are normally called electric and electronic equipment (EEE). EEEs may have a short application period and some with short lifespan, and are often discarded in landfills after usage or bundled in store rooms. While waste management is a global challenge, electronic waste (e-waste) has added pain because it consists of dangerous assortment of toxic metals and plastics. This has prompted many countries including South Africa to classify e-waste as hazardous waste that need to be checked at all times. Nonetheless, recent research show potential for reuse as high energy rich materials. In fact, it could be reused, remanufactured, recycled, landfilled and/or incinerated. There are alternative thermal processing techniques such as pyrolysis and gasification instead of incineration. This chapter covers global perspective on the regulations of e-waste in selected countries including South Africa, the types of e-wastes, the recycling of e-waste, environmental issues related to e-waste and the thermal processing of e-waste into biofuels.

**Keywords** Electronic equipment · Governmental regulations · Recycling

## 1 Introduction

Worldwide, electrical and electronics industry is growing at a fast rate due to the development of new technologies related to electronics, which increases the demand of electrical and electronic devices [1, 2]. Electrical and electronic devices are termed electrical and electronic equipment (EEE), thus referring to any device that operates using electricity as the main source of power. These include complex machines that are used daily such as robots, computers, automatic transfer machines, different types of sensors and gadgets to say the least [3]. At industrial level the production gives

T. E. Motaung (✉)
Department of Chemistry, Sefako Makgatho Health Sciences University, P.O. Box 94, Medunsa 0204, South Africa
e-mail: tshwafo.motaung@smu.ac.za; motaungte@live.com

 
M. Tshwafo Elias (ed.), *Waste to Value Innovations*,
https://doi.org/10.1007/978-981-96-2782-0_1

very diverse e-waste consisting mainly of computer peripherals, laptops, televisions, telephones, cellphones, printers, refrigerators, washing machines, radios, electronic storage servers, telecommunication devices, video games, cameras, video recorders and other range of electrical devices, both household and industrial electronics all part of EEE [4–10]. EEE are widely used, however, the lifespan is short, in which a large portion of EEE ends up as waste. That has rendered EEE industry responsible for the generation of approximately 42 million tons per year of EEE waste (e-waste), which is also referred to as WEEE (waste electrical and electronic equipment), or electronic scrap [3]. At municipal level e-waste only accounts for 1–3% of the produced municipal waste, nonetheless globally this count and expands the problem [6–8].

E-waste could be classified as hazardous waste which needs to be handled properly since it has a high potential of causing health related issues, environmental and ecological problems [9]. Furthermore, the discarding of electronics has given rise to environmental concerns as the e-waste is toxic to the population and environment [1]. In fact, it is amongst the fastest growing waste streams in Southern Africa. Possibly due to increasing market penetration from developing countries and growing rate of electronic imports to advance the recent competitive global technology devices [2, 10]. Its management, through ordinary recycling, is a challenge because of its inherent composites structure. It is more challenging when disposed in a landfill because the landfill, in most cases, is accompanied by other forms of wastes. As a results the separation of wastes into individual manageable components becomes a nightmare [8, 10]. For instance, the recycling of electronic scrap commonly depends on pyrometallurgical treatment in copper smelters. But the treatment of electronic scrap with high amount of plastics and other wastes needs constantly a combination of mechanical, thermal and hydrometallurgical, whereas the environmental regulations have to be considered. But the costs of sampling and analysis of base and precious metal scrap are quite high and they are often higher than the economics of processing [4]. That is one of the factors affected revenues in the e-waste of small business enterprises. For example, South Africa has over 100 e-waste recycling registered companies [8, 10]. However, most companies clearly do not do much from e-waste recycling alone. Some resorted for dismantling e-waste into individual rich components that could be transported abroad [10]. Ferrous, non-ferrous, and printed circuit boards (PCBs) are the main outputs of SA e-waste recycling. PCBs are the mostly exported at about 90%, with phosphor powders being exported at 60% and some ferrous and non-ferrous are also exported [10–14].

## *1.1 Regulations of e-Waste*

In terms of the South African Waste Amendment Act (2014), waste means

"*(a) any substance, material or object, that is unwanted, rejected, abandoned, discarded or disposed of, or that is intended or required to be discarded or disposed of, by the holder of that substance, material or object, whether or not such substance, material or object can be re-used, recycled or recovered and includes all wastes as defined in Schedule 3 to this Act; or (b) any other substance, material or object that is not included in Schedule 3 that may be defined as a waste by the Minister by notice in the Gazette, but any waste or portion of waste, referred to in paragraphs (a) and (b), ceases to be a waste—*

(i) *once an application for its re-use, recycling or recovery has been approved or, after such approval, once it is, or has been re-used, recycled or recovered*;
(ii) *where approval is not required, once a waste is, or has been re-used, recycled or recovered*;
(iii) *where the Minister has, in terms of section 74, exempted any waste or a portion of waste generated by a particular process from the definition of waste*; *or*
(iv) *where the Minister has, in the prescribed manner, excluded any waste stream or a portion of a waste stream from the definition of waste*" [8].

WEEE is classified under hazardous waste (CODE HW 18). In terms of the South African Waste Amendment Act (2014) states that hazardous waste means "*any waste that contains organic or inorganic elements or compounds that may, owing to the inherent physical, chemical or toxicological characteristics of that waste, have a detrimental impact on health and the environment and includes hazardous substances, materials or objects within business waste, residue deposits and residue stockpiles*" [8]. European Union (EU) has waste control and management regulations over EU countries, which includes their e-waste regulations. The European Directive defines e-waste as "*waste equipment that is dependent on electric currents or electromagnetic fields to work properly and equip for the generation, transfer, and measurement of such currents*" [14].

Since the amount of generated waste of a country has a linear relationship to the increase of its population, level of income and urbanisation of the country. That is the state which appeared to have contributed in many countries including South Africa and that resulted in disposal to landfills in some industries which have challenges of enforcement and implementation of waste management legislation [8]. Actually, countries such as South African (SA) experience more pain when coming local government in terms of collection and processing the randomly disposed e-waste. There is also a limited capacity in rural areas that ensures effective and efficient waste management for SA. Some studies recommend a change in behaviour and more public awareness in order to overcome some of the well-known challenges [8, 10, 11]. Those problems include:

- Poor waste collection services due to inadequate landfill available to accommodate the growing SA population.
- Plenty of illegal waste dumping sites and unlicensed and unpermitted waste management and disposal activities.
- Limited landfill space available, national landfills are almost full.
- Lack of implementation of current waste control and recycling initiatives.
- Insufficient public information and awareness with regards to waste.
- Waste legislation is not regulated and enforced properly.

# 2 Types of Electronic Waste

## 2.1 Classification and Categories of EEE

Classification of EEE differs for each country depending on their regulations. It is mostly classified as industrial electronics, computer peripherals, lighting devices, household appliances and telecommunication devices [2]. Industrial electronics are automobile electronic devices, medical devices, sensors, etc. Computer peripherals are computer hardware devices such as laptops, monitors, keyboards, circuit boards, mouse, floppy discs, servers, etc. lighting devices are fluorescent tubes. Household appliances are fridges, televisions, radios, washing machines, cameras, video and kettles. Telecommunication devices are cellular phones, phones, fax machines, routers, transmitters, and Radio Frequency equipment (RF) [2, 14]. E-waste has two main categories as per the E-waste Management and Handling Rules, 2011; electrical and electronic equipment and consumer electrical and electronics. EU Directive has categorised EEE in 10 categories as shown in Table 1.

## 2.2 Chemical Composition of e-Waste

E-waste has different categories with different EEE's, which means that the chemical composition is dependent on the EEE type and the lifespan [12]. Recycling of e-waste provides huge amounts of raw materials recovery and the protection of land that would otherwise be claimed as landfills. Its composition depends on the type of materials used in the production of EEEs. In most cases, it is composed of ferrous and non-ferrous metals, pollutants, printed circuit boards, plastics, glass, concrete, ceramics, wood, rubber, and other materials [12, 14]. The main component chiefly iron and steel, which is about 50% of total e-waste. The other components are plastics, which are about 21%, non-ferrous metals, which are about 13%, and the rest of the components make up the remaining 16% [2] (see Fig. 1). The non-ferrous metals include copper, aluminium, silver, gold, platinumpalladium. The other metals found in e-waste are mercury, arsenic, selenium, cadmium, lead, and flame retardants polyvinyl chlorides (PVCs), brominated flame retardants, beryllium, antimony, and phthalates, of which their presence in certain quantities often causes toxicity and are hazardous [2]. Hazardous and non-hazardous materials normally attributed to e waste are summarised in Table 2. In fact, there are more than 1000 various substances found in e-waste, including elements, various metal alloys, and organic pollutants [12, 14].

Amongst municipal waste types, e-waste is the most hazardous due to the toxic chemicals and metals found in the EEE. Human long-term exposure to these materials has severe health effects since some are carcinogenic, and neurotoxic, they are harmful in the nervous system, kidneys, bones, reproductive system [2].

**Table 1** EEE EU directive categories adopted from Letcher and Vallero [14]

| No. | Category | Major electric and electronic equipment |
|---|---|---|
| 1 | Large household appliances | Refrigerators<br>Washing machines<br>Cooking<br>Electric fans, air conditioner appliances |
| 2 | Small household appliances | Vacuum cleaners<br>Irons<br>Toasters, grinders, coffee machines<br>Electric knives, hair-cutting, tooth brushing<br>Clocks, watches, scales, etc. |
| 3 | IT and telecommunications equipment | Mainframes<br>Data processing, personal computers, laptops, notepads, calculations, Printer units, copying equipment, facsimile, telephones |
| 4 | Consumer | Television sets<br>Radio sets<br>Video cameras, recorders, amplifiers, etc. |
| 5 | Light equipment | Luminaires for fluorescent lamps<br>Straight or compact fluorescent lamps<br>High-intensity discharge lamps<br>Low-pressure sodium lamps |
| 6 | Electrical and Electronic tools | Drills<br>Saws<br>Sewing machines |
| 7 | Toys, leisure, and sports equipment | Electric trains or car racing sets<br>Video games or consoles<br>Computers for biking, diving, running, rowing, etc.<br>Coin slot machines |
| 8 | Medical devices | Radiotherapy equipment<br>Cardiology<br>Dialysis<br>Pulmonary ventilators |
| 9 | Monitoring and control instruments | Smoke detectors<br>Heating regulators<br>Thermostats |
| 10 | Automatic dispensers | Automatic dispensers for hot or cold drinks, solid products, etc. |

## 3 Recycling of Electronic Waste

E-waste is considered hazardous waste in South Africa. In 2017, South Africa generated 68 million tonnes of hazardous waste, of which only 6% was re-used or recycled. The remaining 94% was either treated and/or landfilled [8]. Although it generally accounts for only 3–5% in solid waste, the composition varieties and contents present

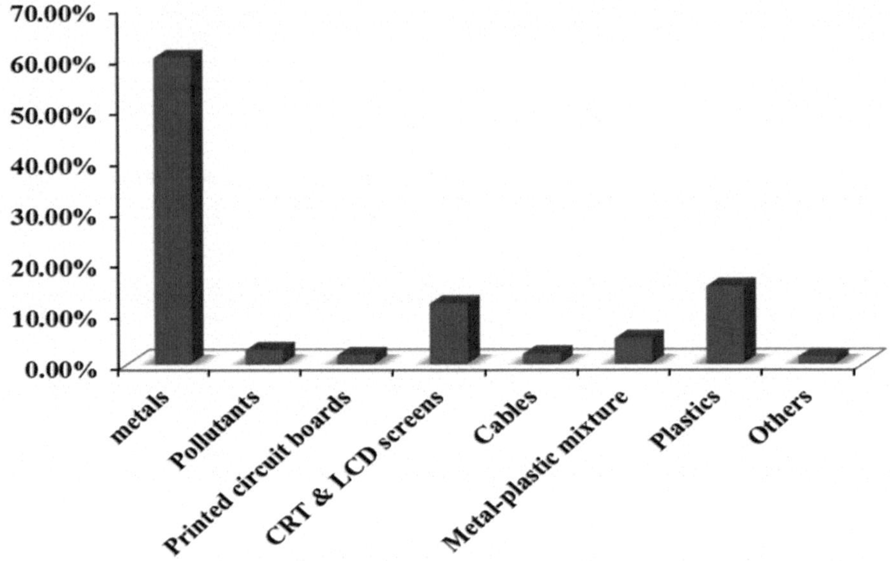


**Fig. 1** Composition of e-waste [13]

in e-waste are far more valuable metals than other municipal waste forms [15]. It, in most cases, forms a very small portion of hazardous waste, with only 0.5% generated (Fig. 2).

The annual global production of e-waste is estimated to between 20 to 50 million tonnes, with only 15–20% being recycled [8, 12]. E-waste can be managed by the reuse of EEE, the remanufacturing of EEE, and recycling. The other alternatives are incineration and landfilling [2, 14]. Reusing end-of-life EEE is often the first priority over the other methods due to the extension of lifespan of the EEE and reduces the generated e-waste being discarded in landfills [14]. Remanufacturing is the second priority in e-waste management, where used EEE products are disassembled, cleaned, repaired, and reassembled, and restored as used as new equipment or addition to new equipment [16]. Remanufacturing retains energy, and the material which is beneficial to the environment as the reused electronic equipment do not end up in landfills.

Recycling EEE is the reprocessing of e-waste materials for several uses as valuable materials [14]. The other benefit of recycling is the energy savings compared to using raw, virgin materials. In e-waste recycling, there are four processes used, namely; fragmentation, size reduction, homogenization, and metals reclamation [12]. Fragmentation of e-waste involves the dismantling of EEE which results in ferrous and non-ferrous metals, plastics, glass, and other elements [12] (Fig. 3).

E-waste recycling can be divided into three major steps under the four processes [14];

**Table 2** Major hazardous components found in e-waste adopted from Letcher and Vallero [14]

| Materials and components | Descriptions |
|---|---|
| Batteries | A large proportion of heavy metals such as lead mercury a d cadmium is present in batteries |
| Cathode ray tubes (CRTs) | Lead in the cone glass and fluorescent coating over the inside of panel glass |
| Mercury containing components, such as switches | Mercury is basically used in thermostats, sensors, relays, and switches (e.g., on Printed Circuit Boards and in measuring equipment and discharge lamps). Furthermore, it is used in medical equipment, data transmission, telecommunication, and mobile phones |
| Asbestos waste | Asbestos waste has to be treated selectively |
| Toner Cartridges, liquid and pasty, as well as colour toner | Toner and toner cartridges may contain in hazardous chemicals that are moderately toxic if acute exposure occurs |
| Printed Circuit Boards | In Printed Circuit Boards, lead, and cadmium are common in solder, cadmium occurs in certain components, such as SMD chip resistors, infrared detectors, and semiconductors |
| Polychlorinated biphenyl (PCB) containing capacitors | PCB-containing capacitors have to be removed for safe destruction |
| Liquid Crystal Displays (LCDs) | Lead and mercury occur in LCDs, especially for old designs LCDs of a surface greater than 100 $cm^2$ have to be removed from WEEE |
| Plastics containing halogenated flame retardants | During incineration/combustion of the plastics, halogenated flame retardants can produce toxic components |
| Equipment containing CRC HCFC or HFCs | The CFC present in the form and the refrigerating circuit must be properly extracted and destroyed; HCFC or CFC's present in the foam and refrigerating circuit must be properly extracted and destroyed or recycled |
| Gas discharge lamps | The mercury has to be removed |

1. **Disassembly step**: is the crucial step in fragmentation, it is the process of selectively dismantle e-waste by targeting the hazardous and non-hazardous components, and valuable components of e-waste.
2. **Upgrading step**: process of material preparation for refining process by mechanical processing and/or metallurgical processing.
3. **Refining step**: process of retreating or purifying recovered materials for their reuse in the production of EEE.

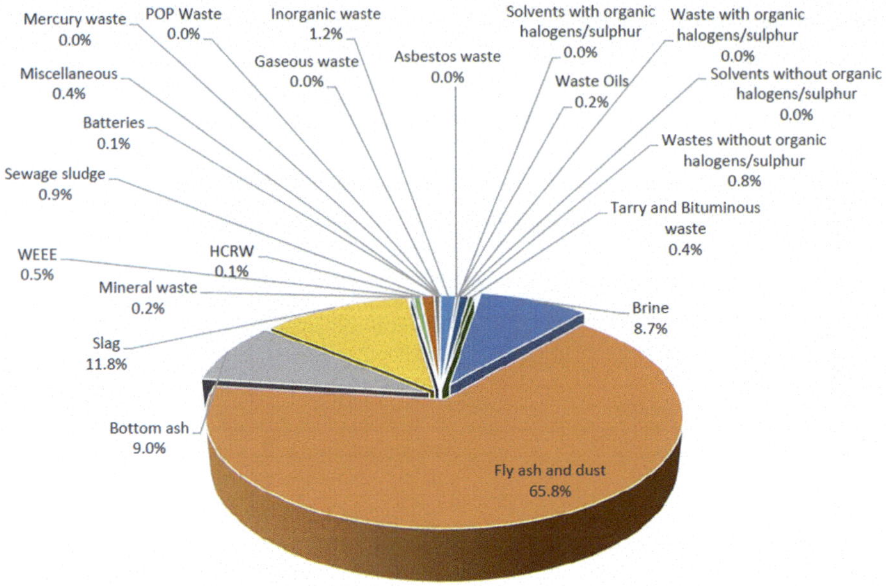


**Fig. 2** Composition of hazardous waste generated annually in SA [8]

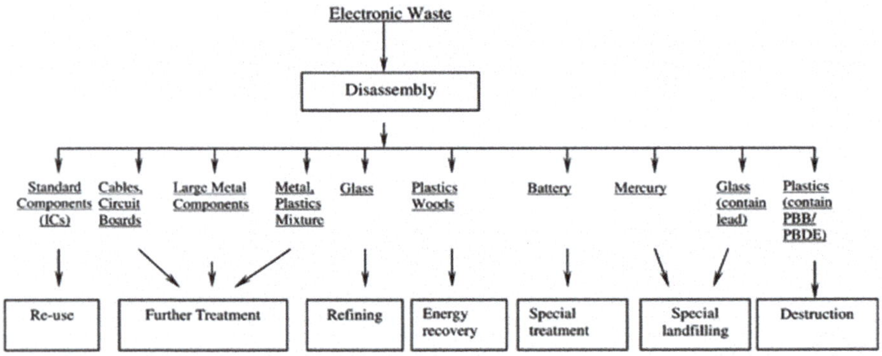


**Fig. 3** Recycling e-waste process adopted from [14]

In terms of recovering precious metals, mechanical processing in insufficient. Thus the recovered metals are further processed by metallurgical and chemical techniques, such as pyrometallurgical and hydrometallurgical processes [15–17]. The last alternatives of treating e-waste is incineration and dumping waste in landfills. E-waste recycling does have a disadvantage as it increases the amount of pollutants exposed in soil, plants, air, water and sediments due to the presence of toxic chemicals [13].

## 4 Environmental and Health Issues of e-Waste

The hazardous chemicals present in e-waste highly impacts the environment and also causes health related issues in humans residing in that area [2]. The toxicity of metals in e-waste results in public exposure through water, air, food, consumer products and agricultural products [18]. E-waste treatment mainly recycling, which involves dismantling, shredding, and separation of e-waste, metallurgical and chemical processing techniques, results in the release of toxic substances to the environment [15]. Due to the several materials present in EEEs, like metals and plastics, handling e-waste results in exposure to the highly toxic mixtures of chemicals present which is detrimental to food chain and ecosystem as summarised in Fig. 4 [19].

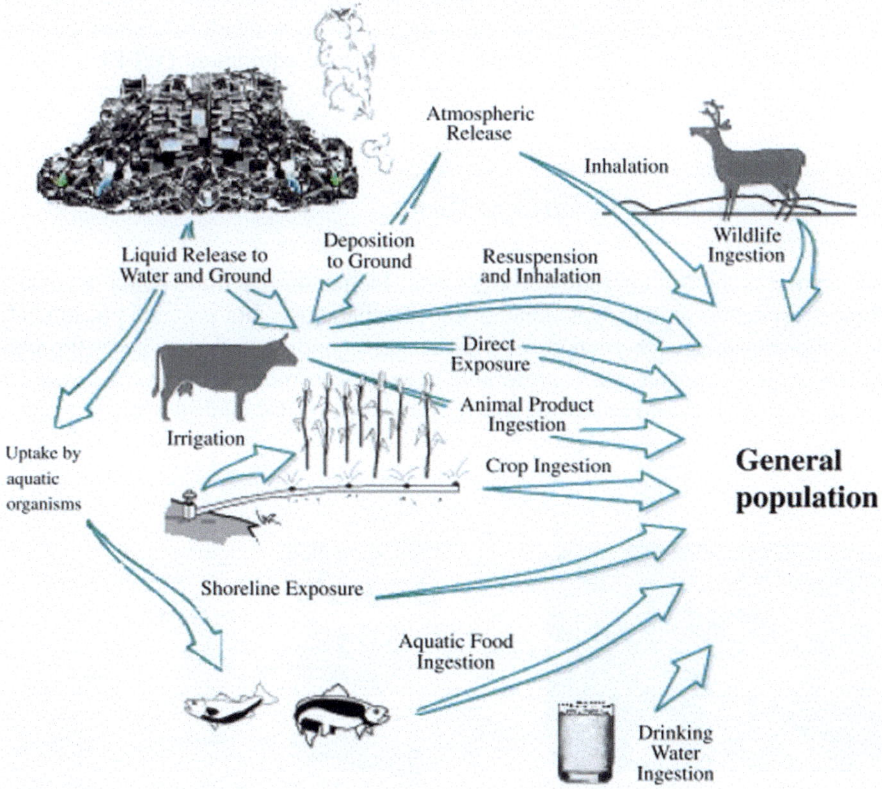


**Fig. 4** Toxicants routes in their exposure in the environment and their contamination in the food chain adopted from Gurgul et al. [19]

## 5 Thermal Conversion of Electronic Waste into Energy

Constructional complexity of items and their integration are the most distinctive features of electronic wastes. These wastes consist of mineral and polymeric materials and have high content of valuable metals that could be recovered. Elimination of polymeric components (especially epoxy resins) while leaving non-volatile mineral and metallic phases is the purpose of thermal treatment of electronic wastes (see Fig. 5). In the case of gasification, gaseous product of the process may be, after cleaning, used for energy recovery or chemical synthesis. If not melted, metals from solid products of thermal treatment of electronic waste could be recovered by hydrometallurgical processing. Three basic, high temperature ways of electronic waste processing, i.e., smelting/incineration, pyrolysis and steam gasification [20].

Rapid economic growth has resulted in an increase of production of electrical and electronic equipment (EEE) which is one of the fastest growing global manufacturing and it is still developing. Waste electric and electronic equipment (WEEE) is now considered as one of the fastest growing waste streams.

### 5.1 *Incineration of Electronic Waste*

Incineration is the last alternative for e-waste. Incineration is done under pyrometallurgical processing, which is done at high temperatures in a gaseous phase [21]. Incineration is an open thermal process which has a continuous flow of gases through the whole process, thus not a preferable e-waste treatment technique because of terrible air pollution and undesired ashes, in most cases, for landfill [20].

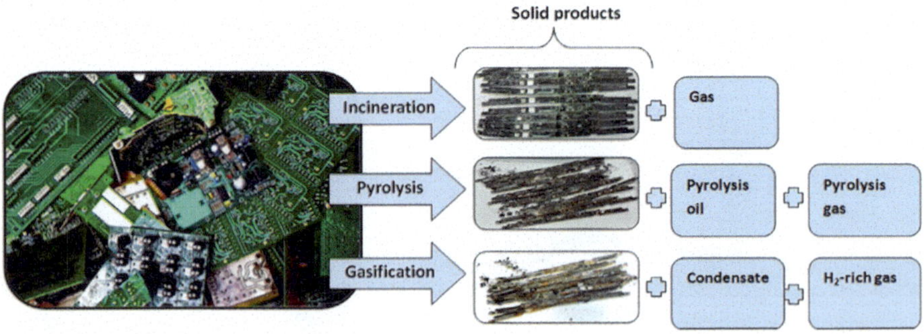


**Fig. 5** Thermal processes of e-waste and their gaseous and liquid products [20]

### 5.2 *Pyrolysis of Electronic Waste*

Pyrolysis is advantageous over incineration as a promising route for the recycling of heterogeneous, contaminated and additives containing plastics from e-waste [20, 21]. It is a thermal processing technique that decomposes organic material at high temperatures without the presence of oxygen. The organic materials change their chemical composition and decomposes to renewable energy fuel. The end products of pyrolysis are gaseous and liquid fuel, with char as a carbon-rich solid residue [20]. Pyrolysis process requires only temperature for the generation of gaseous products [21]. That is in fact, circular economy at its best because obviously the final products (Biofuels, tar and oil) have additional value for a market. However, pyrolysis also produces char, which hinders the process of recovering metals during e-waste treatment.

### 5.3 *Gasification of Electronic Waste*

E-waste can be processed by a process called gasification [20]. Gasification is often grouped with pyrolysis as they both generate biofuels through high temperature processing. However, unlike pyrolysis which only requires high temperature for the production of biofuels, gasification requires an additional agent [21]. In gasification process, there is a complete maximum energy transfer, which converts organic materials into syngas [20]. Furthermore, gasification uses gasifying reagents, namely; steam, oxygen, air, oxygen enriched air or carbon dioxide. Steam gasification is an endothermic process which is referred to as allothermal, while the other gasification methods are referred to as autothermal [21]. Steam gasification is the preferred method of e-waste treatment compared to the other gasification methods as the organic materials do not hinder the recovery of metals as maximum energy is transferred during the process [20].

## 6 Conclusion

The electronics industry is one of the most important industries in the world. It has grown steadily in recent decades, and will be growing more in the next decades due to demands by fourth industrial revolution. It also generates a high employment rate, improves technology development. Electrical and electronic equipment is on high demand, which is matched by the high production of EEE. However, the disposal of e-waste remains a global problem. There are companies available which are monitoring e-waste globally and in SA. Incineration, pyrolysis and gasification are used to convert e-waste into energy products. More work needs to be done in the conversion of e-waste into energy as most companies are interested in the recycling of metals. The handling of e-waste is also a major concern due to e-waste consisting of hazardous components.

## References


1. Xianlai, Z., Qingbin, S., Jinhui, L., Wenyi, Y., Huabo, D., & Lili, L. (2015). Solving e-waste problem using an integrated mobile recycling plant. *Journal of Cleaner Production, 90*, 55–59.
2. Rao, M. N., Sultana, R., Kota, S. H., Shah, A., & Davergave, N. (2016). *Solid and hazardous waste management: Science and engineering*. Butterworth-Heinemann.
3. Bindschedler, S., Bouquet, T. Q. T. V., Job, D., Joseph, E., & Junier, P. (2017). Fungal biorecovery of gold from e-waste. *Advances in applied microbiology, Elsevier., 99*, 53–81.
4. Gramatyka, P., Nowosielski, R., & Sakiewicz, P. (2007). Recycling of waste electrical and electronic equipment. *Journal of Achievements in Materials and Manufacturing Engineering, 20*, 535–538.
5. Sodhi, M. S., & Reimer, B. (2001). Models for recycling electronics end-of-life products. *OR-Spektrum, 23*(1), 97–115.
6. Goodship, V., & Stevels, A. (2012). *Waste electrical and electronic equipment (WEEE) handbook*. Elsevier.
7. Godfrey, L., & Oelofse, S. (2017). Historical review of waste management and recycling in South Africa. *Resources, 6*(4), 57.
8. Department of Environmental Affairs. (2018). South Africa State of Waste. A report on the state of the environment. First draft report. Department of Environmental Affairs. Pretoria.
9. Song, Q., & Li, J. (2014). A systematic review of the human body burden of e-waste exposure in China. *Environment international, 68*, 82–93.
10. Lydall, M., Nyanjowa, W., & James, Y. (2017). Mapping South Africa's waste electrical and electronic equipment (WEEE) dismantling. In *Pre-processing and processing technology landscape* (pp. 1–80). Waste Research Development and Innovation Roadmap Research Report, Mintek. Randburg, South Africa: Mintek.
11. Nahman, A., & Godfrey, L. (2010). Economic instruments for solid waste management in South Africa: Opportunities and constraints. *Resources, Conservation and Recycling, 54*(8), 521–531.
12. Tzoraki, O., & Lasithiotakis, M. (2018). Environmental risks associated with waste electrical and electronic equipment recycling plants. In *Reference module in earth systems and environmental sciences*. Elsevier.
13. Zhang, L., & Xu, Z. (2016). A review of current progress of recycling technologies for metals from waste electrical and electronic equipment. *Journal of Cleaner Production, 127*, 19–36.
14. Letcher, T. M., & Vallero, D. A. (2011). *Waste: A handbook for management*. Academic Press.
15. Li, J., Zeng, X., Chen, M., Ogunseitan, O. A., & Stevels, A. (2015). "Control-Alt-Delete": Rebooting solutions for the e-waste problem. *Environmental Science & Technology, 49*(12), 7095–7108.
16. Williams, J., & Shu, L. (2001). Analysis of remanufacturer waste streams for electronic products. In *Proceedings of the 2001 IEEE International Symposium on Electronics and the Environment*. 2001 IEEE ISEE (Cat. No. 01CH37190), IEEE.
17. Nordberg, G. F., Fowler, B. A., & Nordberg, M. (2014). *Handbook on the toxicology of metals*. Academic press.
18. Frazzoli, C., Orisakwe, O. E., Dragone, R., & Mantovani, A. (2010). Diagnostic health risk assessment of electronic waste on the general population in developing countries' scenarios. *Environmental Impact Assessment Review, 30*(6), 388–399.
19. Gurgul, A., Szczepaniak, W., & Zabłocka-Malicka, M. (2018). Incineration and pyrolysis vs. steam gasification of electronic waste. *Science of The Total Environment, 624*, 1119–1124.

20. Cui, J., & Zhang, L. (2008). Metallurgical recovery of metals from electronic waste: A review. *Journal of Hazardous Materials, 158*(2), 228–256.
21. Santella, C., Cafiero, L., De Angelis, D., La Marca, F., Tuffi, R., & Ciprioti, S. V. (2016). Thermal and catalytic pyrolysis of a mixture of plastics from small waste electrical and electronic equipment (WEEE). *Waste Management, 54*, 143–152.

# Thermal Conversion of Solid Wastes

**Tshwafo E. Motaung, Lehlohonolo F. Koao, and Setumo V. Motloung**

**Abstract** Solid waste management is a serious environmental problem which is directly linked to the amount of accumulated waste production and shortage of waste disposal areas. Solid waste is piling up in landfills and causing uncontrollable greenhouse gases emission. Apart from land pollution, it also causes water and soil pollution and increases health risks. The goal is to reduce the amount of solid waste that is dumped in the environment. One of the most effective techniques is the thermal conversion of waste into energy sources. This chapter looks at the thermal conversion techniques that has a potential of converting solid waste into biofuels. Pyrolysis, gasification and combustion are all alternatives to incineration as they result in clean energy sources. Solid waste is available to be utilised into energy as an alternative method to the use of fossil fuels.

**Keywords** Solid waste · Greenhouse gases emission · Thermal treatments

## 1 Introduction

### *1.1 Background*

Globally, solid waste is generated at higher rates that causes environmental problems, which fill up landfills with less treatment techniques available to minimise waste. The utilisation of biomass as raw materials in energy generation is needed, as well

T. E. Motaung (✉)
Department of Chemistry, Sefako Makgatho Health Sciences University, P.O. Box 94, Medunsa 0204, South Africa
e-mail: tshwafo.motaung@smu.ac.za; motaungte@live.com

L. F. Koao
Department of Physics, University of the Free State (Qwa Qwa Campus), Private Bag X13, Phuthaditjhaba 9866, South Africa

S. V. Motloung
Central University of Technology, Private Bag X20539, Bloemfontein 9300, South Africa

 
M. Tshwafo Elias (ed.), *Waste to Value Innovations*,
https://doi.org/10.1007/978-981-96-2782-0_2

as reusing solid waste for bio-based products [1]. The conversion of solid waste into renewable energy is also beneficial in the reduction of greenhouse gases (GHGs) emission in a country like South Africa, which produces the highest amount of GHGs compared to other African countries [2]. Innovative uses of biomass into renewable energy is necessary, which is also beneficial in the reduction of solid waste in landfills. One of the integrated solid waste management systems is the thermal conversion of solid wastes and biomass into sustainable renewable energy [1].

In previous years, the reduction of solid waste has been managed by recycling, waste combustion and disposing it in landfills as waste management hierarchy which was incomplete. The completion of waste management hierarchy involves chemical, biological or thermal treatment of waste [3]. Thermal conversion of waste reduces waste occupying landfills by converting waste into harmless products by using heat, steam, electrical energy and combustible energy all found within the solid waste [3, 4]. In every country, waste management which involves the treatment, management and disposal of solid waste in a requirement.

## 2 Biomass

Biomass is plant-based and animal-based materials of living, or recently living, or dead organic matter, which are all renewable source of energy due to the release of heat energy when combusted or undergoing decomposition [5]. Biomass includes hardwood and softwood (forest residues); agricultural residues (wheat straw, corn cob, coconut husks, wheat rice, bagasse, etc.); industrial residues and municipal solid waste (MSW); energy crops and food wastes [6, 7]. The disposal of biomass waste can be harmful to the environment, by either piling up landfills and causing pollution or emitting GHGs when burnt up, which has a huge impact on the environment and the ecosystem. Thermal treatment of solid biomass waste can result in activated carbons and renewable energy sources [8]. MSW includes food waste, newspapers, plastic used products, wood, textiles, glass etc. from households, industries, schools and hospitals [9].

## 3 Thermal Treatment Techniques for Solid Waste Management

Thermal conversion of solid waste can be achieved in using thermal conversion techniques, torrefaction, pyrolysis and gasification. These methods resulting products are carbon concentrated solid fuels, liquid hydrocarbons, and gaseous fuels [10]. The products obtained from thermal conversion of solid waste and biomass can be utilized as renewable source of energy [11]. Moreover, these techniques can be alternatives of waste management instead of disposal of waste in landfills or incineration.

Thermochemical conversion technologies for the sustainable integration of MSW mainly include gasification, pyrolysis, and torrefaction. These processes are generally characterized by elevated temperatures and fast conversion rates in comparison with biochemical treatment processes. Gasification, pyrolysis, and torrefaction techniques are done in the absence of oxygen or with significantly less oxygen than the required oxygen for complete combustion as shown in Fig. 1 [12].

The operation conditions of these techniques vary with respect to the product yield. The conditions which are controlled are temperature, heating rate, and oxygen supply. In torrefaction, the low temperatures and log residence times produces chars. In pyrolysis, the fast-heating rates and moderate temperature produces liquid products. In gasification, the high heating rates and temperatures produces gaseous products [12]. These techniques cannot be separated completely due to torrefaction being the initial stage of gasification and pyrolysis an incomplete gasification process. Combustion is also considered a thermal conversion process which produces heat and power.

Thermal conversion processes use higher temperatures to effectively treat different types of solid waste better than biochemical and physicochemical processes [13]. Thermal treatment techniques have several advantages compared to the other waste treatment techniques. The advantages are:

- About 70–80% reduction of solid waste in mass and 80–90% of solid waste in volume, freeing landfills for more waste disposal [14].
- Minimising usage of land to dispose waste [15].
- Thermal destruction of halogenated hydrocarbons and safe disposal of inorganic contaminants [16].

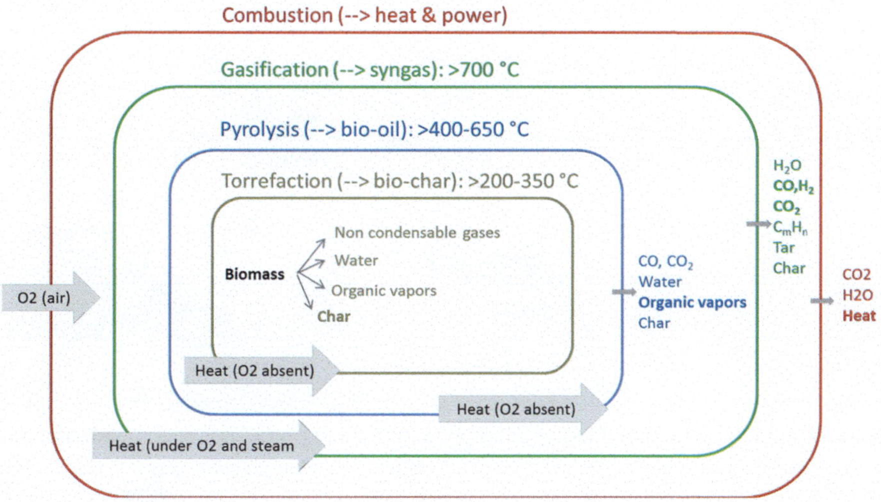


**Fig. 1** Comparison of torrefaction, pyrolysis, gasification and combustion temperature ranges with resulting products [12]

- Thermal residues from the ash and slag are recycled into ferrous and non-ferrous metals [17].
- Combustion of solid waste reduces GHGs emission [15].
- Life cycle assessment studies, which are a standardized environmental management methods concluded that producing renewable energy from solid waste is more beneficial to the environment compared to alternative techniques of different solid waste management [13, 18].

Thermal conversion treatment plants convert the energy value of MSW into different energy forms, e.g., electricity, and that energy is used to operate the plant. One or more thermal conversion technique can be used to power the plants [13].

## *3.1 Pyrolysis*

MSW pyrolysis is considered as an innovative alternative for treating MSW that obtains different chemicals and fuels [4]. MSW treatment has been and will continue to be a major issue facing countries worldwide, and gasification has recently gained more attention as an efficient and clean technology [19]. Pyrolysis is a cleaner technique of producing renewable energy compared to incineration. There is a reduction of amounts of nitrogen oxides and sulphur oxides. There are also less gaseous emissions, with a high-quality solid residues production [20] (Fig. 2).

Pyrolysis has a potential to be used as an alternative to incineration for the reduction of MSW as it results in energy sources [4]. During pyrolysis, there are factors that are taken into consideration, like the temperature, heating rate, and residence time.

**Fig. 2** Pyrolysis process of conversion of MSW into products [21]

These factors are related to the products and the yield of desired products. These are also important in scaling up pyrolysis process for industry. Pyrolysis process is an effective thermal conversion technique of waste-to-energy, however, there are contaminants in the products, such as HCl, $SO_2$ and $NH_3$, and heavy metals which has been reviewed [4]. During pyrolysis, solid waste is thermally degraded in absence of oxygen. The resulting products are char, oil/wax, and combustible gases. For a thousand years, this technique has been used to produce charcoal from biomass. When applied in solid waste, it can produce fuel in a form of gaseous or liquid products, char, and metals [4]. Pyrolysis can be scaled up in a more flexible way compared to incineration and can be a solution to the landfill and incineration issues worldwide, while producing quality energy products. The other important factor to consider for pyrolysis process is the pre-treatment of solid waste. Pre-treatment of solid waste separates the undesirable materials and reduces the size of solid materials before pyrolysis. Moisture is also reduced [4].

MSW has been converted into bio-oil using pyrolysis technique [21]. Bio-oil can be used as a transportation fuel in diesel engines and in gas turbines to generate electricity. It is also less expensive to store and transport, and easier to handle and burn compared to fossil fuels. The activated carbon char (produced as a solid residue during pyrolysis), can also be used as a membrane in water treatment or in agriculture, as a soil amendment [22, 23]. Apart from MSW and biomass, pyrolysis has also been used in industrial waste such as plastics [24] to yield monomers. Pyrolysis has also been used in tyres to produce gaseous products. i.e., $H_2$ and $CH_4$, and aliphatic hydrocarbons and aromatic hydrocarbons [25]. Pyrolysis has also been used in sewage sludge, which resulted in the yield of products such as gas, oil, pyrogenic water, and solid residue (pyrolytic char) [26]. The widely used pyrolysis method is thermal pyrolysis. There is also plasma and microwave pyrolysis. Plasma pyrolysis technique is operated at very high temperatures of over 1000 °C using plasma torches and limits the availability of air. The waste is converted into syngas and other end-products [4]. This method is suitable for the production of energy and recycling materials, as it is also advantageous in being harmless in the environment and public health. However, it requires large amounts of secondary energy and too expensive. Microwave pyrolysis uses homogenous waste such as tyres, shredded plastics and sludge to produce energy. It is a rapid, easily controllable, efficient technique that uses microwave energy for energy production. However, the feedstock particles have to be very fine for this method to achieve a rapid heating rate and the microwave frequency range has to be accurate for achieving the right temperature control [4]. This method is impractical for industrial usage.

## 3.2 *Gasification of Solid Waste and Biomass*

The process whereby solid waste is converted into fuel or synthetic gaseous products is known as gasification, or indirect combustion [13]. It is a partial oxidation of solid waste, where an oxidant present is lower, in concentration, than the amount required for stoichiometric combustion. IN air gasification, the partially combusted fuel provides heat required for gasification, which is known as auto-thermal gasification. In plasma torch utilization, heat is provided by an external source, which is known as allo-thermal gasification [13]. Both of these gasification processes result incompletely oxidised products that have calorific values, which are syngas. Syngas can be separated in different sites or times in the separation process equipment. The products of gasification are carbon monoxide, hydrogen, and lower amounts of methane. Syngas is cleaned up due to the presence of contaminants such as tar, alkali metals, chloride, and sulphide [13].

The following are the properties of gasification of solid waste [11]:

- Gasification has a fast waste reduction rate due to the temperature used.
- Gasification of solid waste minimises the amount of land/landfill sites required for the disposal of waste.
- When there is less solid waste disposed in landfills, there is also fewer toxic pollutants leaching into the soil and polluting ground water. Less soil and water pollutions, and land pollution.

Biomass gasification is used to produce hydrogen. The advantages of using gasification for hydrogen production are as follows [27]:

- Gasification is a fast process.
- Gasification is an efficient process.
- Gasification of biomass is necessary since biomass is environmentally friendly, renewable natural resource and abundantly available.

Gasification process requires a gasification agent for a successful gasification process. Gasification agents are air for air gasification, oxygen for oxygen gasification, steam for steam gasification, etc. these are the widely used gasification agents [27].

Amongst these gasification processes, oxygen gasification and steam gasification yield better gasification results, due to the $N_2$ dilution being avoided during both processes. While both processes perform better than air gasification, steam gasification outperforms oxygen gasification. The oxygen gasification process consumes the combustible gases formed, thus lowering the syngas heating value. While during steam gasification process, the steam produces more combustible gases in the syngas, thus resulting in an increased heating value of syngas [27].

Gasification process can be performed in various types of gasifiers, namely, fixed bed updraft, downdraft gasifier, fluidized bed gasifier and entrained flow gasifier. Downdraft gasifier is shown in Fig. 3.

Syngas produced during gasification can be used as a gaseous fuel in burners or gas engines and produce electricity [28]. It is also used to produce energy products such as chemicals and other fuel energy products [28]. Solid waste is a feedstock for the production of biofuels and biochemicals. Gasification is environmentally advantageous compared to other techniques. Zhou et al. [29] studied the gasification of biomass and quantified the yield of the tar, carbon products, and the yield of $H_2$. Zhang et al. [30] also studied the gasification of biomass for the production of

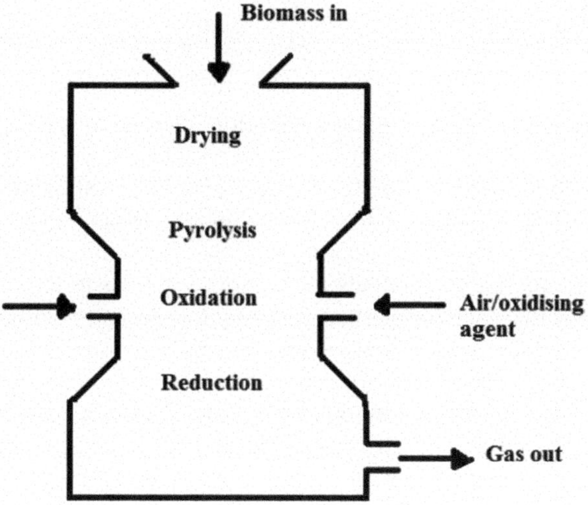


**Fig. 3** Downdraft gasifier [11]

hydrogen. While Irfan et al. [31] recently produced hydrogen through another gasification process called supercritical water gasification. It was noted that the feedstock does affect the hydrogen production yield.

The gasification of MSW in the presence of waste marble powder as a calcium-based catalyst was studied [31]. The reaction resulted in a hydrogen enriched syngas, $CO_2$ adsorption and tar yield. Waste marble powder was able to increase the hydrogen yield while decreasing $CO_2$ [31]. Co-gasification of pine dust and MSW was studied by Cao et al. [32]. The results yielded an increase in CO and $CH_4$, and a decrease in $H_2$ and $CO_2$. Various lignocellulosic biomass was processed during air-steam gasification to produce combustible gases ($H_2$, CO, $CH_4$) [33]. The production of $H_2$ was favoured by higher temperatures and a higher biomass conversion rate. The hydrogen produced was in the range of 30–39% (vol) and combustible gases in the range of 60–78% (vol).

### *3.3 Incineration of Solid Waste*

Incineration is considered a method of minimising solid waste. However, the incineration process could be a source of toxic emissions to the environment. Airborne pollutants have been found to be emitted during incineration [34]. Mercury has also been studied as a possibility to be released in air, soil and groundwater during incineration, thus polluting the environment and causing health issues [35]. There is also a health risk of selected heavy metals, dioxins and volatile organic compounds that are emitted in the atmosphere during incineration [36].

## 4 Conclusion

Due to the environmental concerns associated with burning of fossil fuels, more attention has been paid to develop alternative renewable energy sources in recent years. Among all options for renewable energy utilization, biomass pyrolysis and gasification have been considered over incineration due to their environmental benefits and economic issues. The treatment, management and disposal of solid waste is a common concern in every country. Thermal conversion of solid waste techniques are alternative methods for treating solid waste into producing different chemicals and fuels. These thermal conversion techniques produce biofuels in a cleaner way, emits fewer toxic compounds and results in quality solid products. They also operate in less energy capacity compared to incinerators. They are also much more flexible compared to incinerators. They can reduce environmental pollution while reducing GHGs. This will also reduce the amount of solid waste in landfills. The solid waste is abundantly available, and utilisation of solid waste has benefits over the usage of fossil fuels for energy.

## References

1. Puna, J. F., & Santos, M. T. (2010). *Thermal conversion technologies for solid wastes: A new way to produce sustainable energy*. Waste Management, IntechOpen.
2. Jain, S., & Jain, P. K. (2017). The rise of renewable energy implementation in South Africa. *Energy Procedia, 143*, 721–726. (World Engineers Summit–Applied Energy Symposium & Forum: Low Carbon Cities & Urban Energy Joint Conference, WES-CUE 2017, 19–21 July 2017, Singapore)
3. Moustakas, K., & Loizidou, M. (2010). *Solid waste management through the application of thermal methods*. Waste Management, InTech.
4. Chen, D., Yin, L., Wang, H., & He, P. (2014). Pyrolysis technologies for municipal solid waste: A review. *Waste Management, 34*(12), 2466–2486.
5. Mohlala, L. M., Bodunrin, M. O., Awosusi, A. A., Daramola, M. O., Cele, N. P., & Olubambi, P. A. (2016). Beneficiation of corncob and sugarcane bagasse for energy generation and materials development in Nigeria and South Africa: A short overview. *Alexandria Engineering Journal, 55*(3), 3025–3036.
6. Lee, H., Hamid, S., & Zain, S. (2014). Conversion of lignocellulosic biomass to nanocellulose: structure and chemical process. *The Scientific World Journal, 2014*.
7. Akhtar, N., Gupta, K., Goyal, D., & Goyal, A. (2016). Recent advances in pretreatment technologies for efficient hydrolysis of lignocellulosic biomass. *Environmental Progress & Sustainable Energy, 35*(2), 489–511.
8. El Saied, M., El Naggar, A. M., & Elhosiny, F. I. (2019). A comprehensive investigation on biomass solid waste conversion to a novel catalyst for hydrothermal production of bio-fuel feedstock.
9. Sipra, A. T., Gao, N., & Sarwar, H. (2018). Municipal solid waste (MSW) pyrolysis for bio-fuel production: A review of effects of MSW components and catalysts.
10. Zhang, J., & Zhang, X. (2019). The thermochemical conversion of biomass into biofuels. In *Biomass, biopolymer-based materials, and bioenergy* (pp. 327–368). Elsevier.
11. Thakare, S., & Nandi, S. (2016). Study on potential of gasification technology for municipal solid waste (MSW) in Pune city. *Energy Procedia, 90*, 509–517.

12. Matsakas, L., Gao, Q., Jansson, S., Rova, U., & Christakopoulos, P. (2017). Green conversion of municipal solid wastes into fuels and chemicals. *Electronic Journal of Biotechnology, 26*, 69–83.
13. Arena, U. (2012). Process and technological aspects of municipal solid waste gasification. A review. *Waste Management, 32*(4), 625–639.
14. Consonni, S., Giugliano, M., & Grosso, M. (2005). Alternative strategies for energy recovery from municipal solid waste: Part B: Emission and cost estimates. *Waste Management, 25*(2), 137–148.
15. Psomopoulos, C. S., Bourka, A., & Themelis, N. J. (2009). Waste-to-energy: A review of the status and benefits in USA. *Waste Management, 29*(5), 1718–1724.
16. Buekens, A., & Cen, K. (2011). Waste incineration, PVC, and dioxins. *Journal of Material Cycles and Waste Management, 13*(3), 190–197.
17. Sommer, M., & Ragossnig, A. (2011). Energy from waste in Europe: An analysis and comparison of the EU 27. *Waste Management & Research, 29*(10_suppl), S69-S77.
18. Arena, U., Mastellone, M. L., & Perugini, F. (2003). The environmental performance of alternative solid waste management options: A life cycle assessment study. *Chemical Engineering Journal, 96*(1), 207–222.
19. Xu, F., Wang, B., Yang, D., Qiao, Y., & Tian, Y. (2018). The steam gasification reactivity and kinetics of municipal solid waste chars derived from rapid pyrolysis. *Waste Management, 80*, 64–72.
20. Saffarzadeh, A., Shimaoka, T., Motomura, Y., & Watanabe, K. (2006). Chemical and mineralogical evaluation of slag products derived from the pyrolysis/melting treatment of MSW. *Waste Management, 26*(12), 1443–1452.
21. Mamad Gandidi, I., Susila, M. D., & Agung Pambudi, N. (2017). Production of valuable pyrolytic oils from mixed Municipal Solid Waste (MSW) in Indonesia using non-isothermal and isothermal experimental. *Case Studies in Thermal Engineering, 10*, 357–361.
22. Van Duck, P. J., & Van De Voorde, H. (1984). Activated charcoal and microflora in water treatment. *Water Research, 18*(11), 1361–1364.
23. Rajapaksha, A. U., Chen, S. S., Tsang, D. C., Zhang, M., Vithanage, M., Mandal, S., et al. (2016). Engineered/designer biochar for contaminant removal/immobilization from soil and water: potential and implication of biochar modification. *Chemosphere, 148*, 276–291.
24. Bockhorn, H., Hornung, A., & Hornung, U. (1998). Stepwise pyrolysis for raw material recovery from plastic waste. *Journal of Analytical and Applied Pyrolysis, 46*(1), 1–13.
25. Xu, S., Lai, D., Zeng, X., Zhang, L., Han, Z., Cheng, J., et al. (2018). Pyrolysis characteristics of waste tire particles in fixed-bed reactor with internals. *Carbon Resources Conversion, 1*(3), 228–237.
26. Gerasimov, G., Khaskhachikh, V., Potapov, O., Dvoskin, G., Kornileva, V., & Dudkina, L. (2019). Pyrolysis of sewage sludge by solid heat carrier. *Waste Management, 87*, 218–227.
27. Zhang, Y., Xu, P., Liang, S., Liu, B., Shuai, Y., & Li, B. (2019). Exergy analysis of hydrogen production from steam gasification of biomass: A review.
28. Seo, Y. -C., Alam, M. T., & Yang, W. -S. (2018). *Gasification of municipal solid waste.* Gasification for Low-grade Feedstock, IntechOpen.
29. Zhou, L., Yang, Z., Tang, A., Huang, H., Wei, D., Yu, E., & Lu, W. (2019). Steam-gasification of biomass with CaO as catalyst for hydrogen-rich syngas production.
30. Zhang, Y., Li, L., Xu, P., Liu, B., Shuai, Y., & Li, B. (2019). Hydrogen production through biomass gasification in supercritical water: A review from exergy aspect.
31. Irfan, M., Li, A., Zhang, L., Wang, M., Chen, C., & Khushk, S. (2019). Production of hydrogen enriched syngas from municipal solid waste gasification with waste marble powder as a catalyst.
32. Cao, Y., Fu, L., & Mofrad, A. (2019). Combined-gasification of biomass and municipal solid waste in a fluidized bed gasifier.
33. González-Vázquez, M. D. P., García, R., Gil, M. V., Pevida, C., & Rubiera, F. (2018). Comparison of the gasification performance of multiple biomass types in a bubbling fluidized bed. *Energy Conversion and Management, 176*, 309–323.

34. Jones, A. M., & Harrison, R. M. (2016). Emission of ultrafine particles from the incineration of municipal solid waste: A review. *Atmospheric Environment, 140*, 519–528.
35. Cheng, H., & Hu, Y. (2011). Mercury in municipal solid waste in China and its control: A review. *Environmental Science & Technology, 46*(2), 593–605.
36. Setyan, A., Patrick, M., & Wang, J. (2017). Very low emissions of airborne particulate pollutants measured from two municipal solid waste incineration plants in Switzerland. *Atmospheric Environment, 166*, 99–109.

# Old Tyres Waste for Carbon Black Films and Pyrolysis Oil

**Thembinkosi D. Malevu, Zolile Mtumela, Benard S. Mwankemwa, and Richard O. Ocaya**

**Abstract** Over the previous few years energy demand has drastically increased that lead to recycling of scrap tyres for the production of fuels for energy in South Africa. This chapter provides a detailed information on how to go about recycling of used tyres into pyrolysis oils and carbon black through pyrolisis method. This method provides clean renewable energy and also solve the difficulty of waste tyre removal in the landfills. During the pyrolysis process wasted tyres are decomposed into low molecular weight compounds under inert thermal conditions. An estimated yields of tyre pyrolysis are typical: 45–55% fuel oil 10–15% steel wire, 30–35% carbon black and 8–10% flammable gas. The recycled tyre pyrolysis oil promises to be a good substitution for diesel fuel and gives a opportunity to minimize the utilization of the natural resources. Due to high specific surface area and electrical conductivity in carborn black it has wide range of appliacation in catalysis, polymers reinforcement and solar energy.

**Keywords** Waste tyre management · Carbon black · Waste tyre pyrolysis products · Energy

T. D. Malevu (✉)
Department of Physical and Chemical Sciences, North-West University (Mafikeng Campus), Mahikeng, South Africa
e-mail: Malevu.td@gmail.com; Thembinkosi.malevu@nwu.ac.za

Z. Mtumela
Department of Chemical and Physical Science, Walter Sisulu University, Mthatha, South Africa

B. S. Mwankemwa
Department of Physics, School of Physical Sciences, College of Natural and Mathematical Sciences, University of Dodoma, Dodoma, Tanzania

R. O. Ocaya
Department of Physics, University of the Free State, Qwaqwa Campus, Phuthaditjhaba, South Africa

 
M. Tshwafo Elias (ed.), *Waste to Value Innovations*,
https://doi.org/10.1007/978-981-96-2782-0_3

# 1 Introduction

The environmental degradation and energy crisis are among the major problems modern society is facing today due to improper disposal of organic solid wastes. The increased amount of organic wastes arise as a results of growing population, industrialization and huge quantities of organic solid waste which are generated by diiferent enterprises daily. The waste tyres are one of the very common and most important hazardous solid wastes disposed everyday all over the world. With the continued increase in the manufacture of automotive cars and trucks, the generation rate of waste tyres is increasing considerably. Apparently there are approximately 4 billion piece tyres which amassed in landfills [1] with 1.5 billion included each year [2, 3].

In south Africa apperently 60 million waste tyres are amassed also, dispersed all through the nation in 2013 with 11 million being included yearly [4]. This would mean that in a century, 170 million waste tyres will be deposited in landfil if current recycling methods are not expanded. The South African statistics futher demonstrate that the quantity of vehicles are expanding at alarming rate which add a salt into the wound of waste tyres. In general tyres are designed to withstand harsh mechanical and and climate conditions, (for example, ozone, light and microscopic organisms) and their complexe structure makes them hard to reuse. [5–7]. This obviously lead to occupation of landfill, but it seems that a productive space is also threatened (Fig. 1). From a economical point of view, no practical method to adequately reuse the waste tyres has been effectively practicesed [8–11]. As of now, re-stepping of these used tyres to expand their life expectancy is boundless [9]. At the point when waste material is deposited in a landfill, it starts to deteriorate and normally emmits landfill gas, which is roughly 50 percent methane and 50% carbon dioxide. Typically, these gases are discharged into the environment, which can prompt negative ecological and air quality effects. Particularly, discarded tyres dumps present a genuine risk to both nature and human wellbeing as they advance the development of irritations and sickness conveying creepy crawlies.

Waste tyres likewise represent a high danger of flames that can be hard to stifle and could have ecological effects because of uncontrolled outflows of conceivably destructive mixes into the air, soil and groundwater [12]. Tyre fires produce dangerous gases which contain cancer-causing and mutagenic synthetic compounds, this makes the waste tyre dumps very unfortunate [13]. Uncontrollable disposing of waste tyres in the landfills make waste tyres to store water which habour mosquitoes and cause different kinds of illnesses which includes malaria and fever to human kind. Landfilling and accumulating are in this manner right now the simplest types of managing waste tyres. In any case, landfilling and accumulating neglect to use the material, energy and synthetic potential in waste tyres. Currently, environmental policies necessitate the waste to be reused and be managed according to the circular economy principles. Therefore, the need of recycling tyres into useful by products is of great interest so as to accord with the circular economy. Thier management is attracting growing interest in both developing and developed countries because of the

**Fig. 1** Bulk tyres deposited in landfill

increasing demand for recycling raw materials into useful byproducts. Disposed tyres can be made into useful fraction by different processes such as thermolysis, pyrolysis and incineration. Incineration and thermolysis are the most widely used technologies for processing waste tyres for energy applications. However, the methods have gain very low social acceptance due to high levels of $CO_2$ emission during combustion of wastes [12]. Pyrolysis is a thermal process that decomposes organic material into low molecular weight compounds under inert conditions [9]. This process split carbonaceous materials at high temperatures ranging from 400–800 °C and in the absence of oxygen. It is an essential and promising process to tackle used tyres into useful products using minimum amount of energy because it produces gas that can be used to heat the oven in which the process takes place. Pyrolysis breaks down the organic part of a tyre into pyrolytic carbon black, liquid fuel and non-condensed mixture of gases (gaseous fraction).

## 2 Physical and Chemical Properties of Tyres

The primary segments in vehicles tyres are normal (natural) rubber (NR) and engineered rubber (ER) which includes styrene butadiene copolymer (SBR) and (butadiene rubber) butyl elastic (BR) together with carbon black (CB), textile, steel and some other addictive's [14, 15]. Most importantly, all of the components associated are 100% recyclable. In Table 1, chemical composition of the used rubber powder is presented. In general, the steel quantity in tyres is about 15%, and it is more important for the heavy trucks tyres. In fact, tyres could be good source of energy and chemicals as summarised in Table 1, therefore a possible way to recover these

**Table 1** Physiochemical characterisation of waste tyre

| Material/element | Mass percentage (%) |
|---|---|
| Rubber | 54 |
| Carbon black | 29 |
| Textile | 2 |
| Oxidize zinc | 1 |
| Sulfur | 1 |
| Additives | 13 |

useful products is essential. This will promote economic value; enhance environmental value and the influence on the global environment through global warming, ozone layer destruction, and depletion of resources.

The properties such as elasticity, non-decomposable, and inert have led to a number of different applications. Up to date, random uses for scrap tyres incorporate a wide scope of utilizations. African countries are partially handling waste tires by redirecting disposed tyres into shoe making as soles (Imbadada, as they call it in IsiZulu), decoration chairs, bumpers in garages, woven entryway mats from scrap tyre strips, crafted works and bite the dust cut items (Figs. 2 and 3). Used tyres are also applied in agriculture for protecting tomatoes, potatoes, cabbages, peppers, or any other veggies from harsh environmental condition such wind as shown in Fig. 3. They are also very useful in landscaping as soil errosion protection in areas subjected to errosions, in shore protection as breakwaters and in docks and hourbours as dock bumpers and ship fenders. In addition to that tyres are used in roads construction as tyres are mixed with asphalt cement forming a material called asphalt rubber [16]. For example, the main material used in designing "shoes" is rubber. There are two types of rubbers that can be recovered for designing of shoes. These are hard and soft part which are mainly inner and outer ring of tyres. The outer ring is usually the part of tyre that has treads and is practically heavier (see Fig. 2). However, through their making of shoes, the middle part is thrown the one mostly in the dumped.

In developed countries, used tyres are usually used to generate electricity and power process steam that is also used to air-condition and warm buildings. It is reported that, calorific value of tyres is comparable to that of coal and crude oil, a characteristic feature that refines the energy recovery potential of waste tyres. Cement, steel and other industries use waste tyres in conjunction with coal as secondary fuel and at the same time researchers are busy on integrating the use of waste tyres with coal in power plants [17]. United States of america (USA) for example, endorsed the use of scrap tyres as an alternative resource to generate electricity, while Germany demonstrated highest scrap tyres energetic recovery [18]. Regardless of all the secondary usefulness scrap tyres have, only 15–20% are taken for reuse while the remaining 75–80% are generally damped into the earth [19]. It is likely estimated that figures to rapidly increase as more and modern cars are manufactured every year and continual increase of car usage.

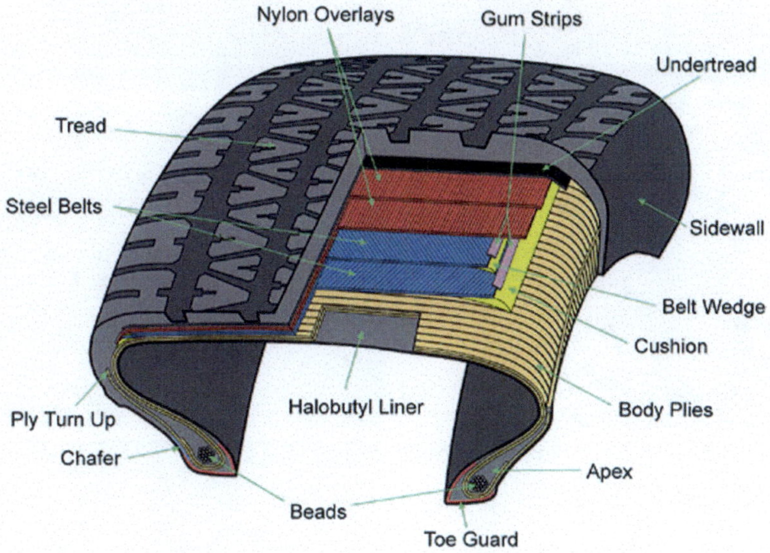


**Fig. 2** The layer construction of a tyre

## 3 Waste Tyre Pyrolysis Products

An alternative way of reducing tyres in landfill is to convert it into activated carbon and other solid carbon forms (carbon black, graphite, and carbon fibres), and liquid fuels (Fig. 4). The key to producing such a product is controlling the chemistry of low temperature carbonization. The South African government recently implemented the Integrated Industry Waste Tyre Management Plan. This means that a large amount of low-quality carbon black (CB) will be produced a material with possessing excellent absorption of electromagnetic radiation, such as sunlight. If this material can be utilised to capture solar energy, it will not only generate clean, renewable energy, but will solve a major waste problem at the same time. Along with CB, oils and syngas will also be produced can be used as fuels or can be further refined into other chemical products.

During the tyre pyrolysis, the principal products are hydrocarbon gases, oils and solid carbon residues (Charcoal). The following yields (as-received basis) of tyre pyrolysis are typical: 33–38 wt.% char, 38–55 wt.% oil, and 10–30 wt.% gas. Figure 5 shows the final product of the waste tyres.

### *3.1 Pyrolysis Oil of Tyres*

The product liquid consists high gross calorific value 40–44 Mj/kg [19]. Which could be enabling their usages as standard fuel, added to overpowering oil generator

**Fig. 3** Secondary uses of tyres

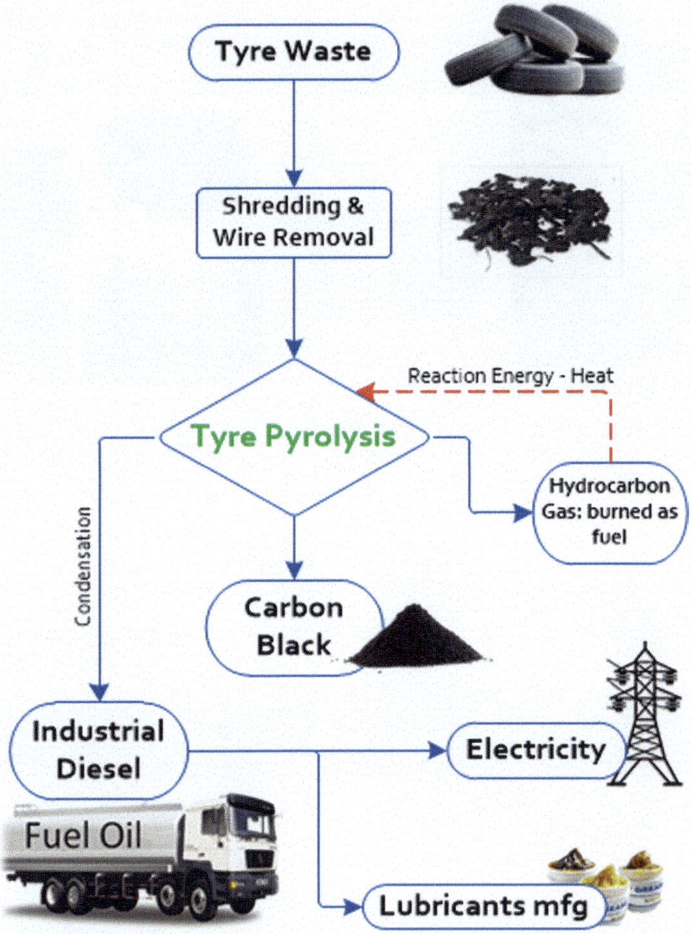


**Fig. 4** Tyre pyrolysis is a process of converting waste tyres into fuel oil

to make control, moved into diesel or gas oil flow equipment. Fuel oil is having high caloric regard and it adds to the business common sense of the waste tyre-reusing plan. However, in some applications with more sophisticated prime movers such as internal combustion engines and aeroderivitive turbines, quality issues with the oil must still be addressed. Upgrading of pyrolysis oil to higher value transportation fuels requires de-oxygenation and reforming of most of the compounds present in the pyrolysis oil. A study conducted by Paasikallio et al. [20] shows that the catalytically produced pyrolysis oil had an average moisture content of 8.3 wt%, and average carbon and oxygen contents of 72.0 and 21.5 wt% as shown in Fig. 6. Because of the large amount of oxygen present, there will be a loss in mass (or volume) yield but this will be balanced against higher heating values.

**Fig. 5** Final product of the waste tyre oil plant

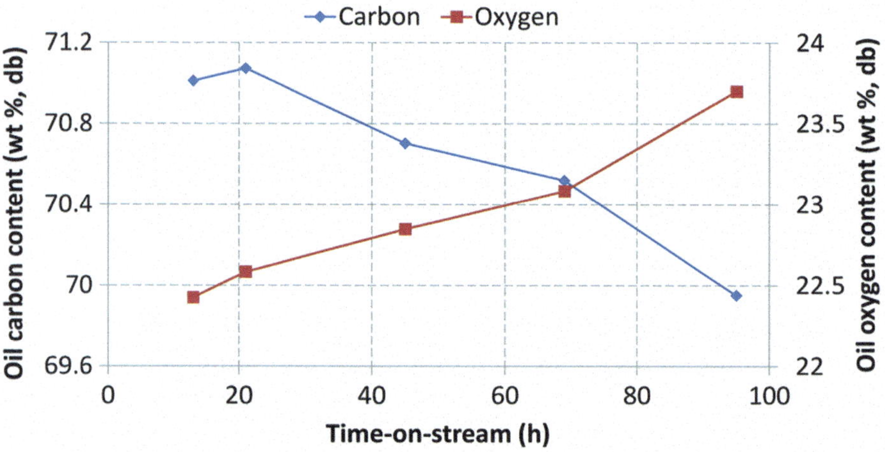


**Fig. 6** Oil fraction carbon and oxygen content (wt%, dry basis) as a function of time-on-stream

Furthermore, during the manufacturing process sulphur is added to the rubber to change the properties of the tyre in a process called vulcanization, whereby at certain temperatures, C–H bonds are broken and replaced sulphur atom chains. The presence of sulphur is undesirable in the petroleum because of their potential corrosion problems in refining process and it is responsible for environmental problems.

The laboratory test results obtained by de Marco Rodriguez et al. [21] shows that physical properties of pyrolysis oil are comparable to that of commercial diesel in medium and high load. Through the process they have successfully reduce the sulphur content of tyre oil to at least 0.04% (from 1044 to 100 ppm) to what has been previously reported by various researchers. That was achieved by incorporating

**Table 2** Comparison of properties of LDO and diesel

| S. No. | Properties | Diesel sample | LDO |
|---|---|---|---|
| 1. | Ash, percent by mass, max | Nil | 0.04% |
| 2. | Flash point | 42 °C | 35 °C |
| 3. | Gross Calorific value | 10800 | 10540 |
| 4. | Viscosity | 2.37 | 4.25 |
| 5. | Sulphur | 0.32<br>0.8 | 0.049 |
| 6. | Water content % by volume (max) | Nil | 0.37 |
| 7. | Cloud point | 5 °C | 7 °C |
| 8. | Sediment % by mass | Nil | Nil |

a mixture of acids such as Hydrogen peroxide ($H_2O_2$), Acetic acid ($CH_2COOH$), Formic acid and sulphuric acids ($H_2SO_2$) to remove sulphur content. Table 2 shows the physical properties of commercial diesel and tyre oil. However, there is still a room of improvement in fuel quality in term of desulfurization; reduction in viscosity is required for tyre oil as an alternate fuel for diesel engine.

Another alternative way of increasing the quality and the yield of pyrolysis oil is to make use of catalyst such as calcium hydroxide, natural zeolite and synthetic zeolite. A Study led by Miandad et al. [22] shows that the use of catalysts not only decreased the overall aromatic hydrocarbons but also formed multiple ring aromatic compounds, except for synthetic zeolite (H-SDUSY) as shown in Fig. 7. Thus, the produced liquid oil has the potential to be used as an alternative source of energy and as a transport fuel, after cleaning and post-treatments.

### 3.2 *Pyrolytic Carbon Black*

Carbon black (CB) is a family of extremely black, finely powder of amorphous carbon with large surface area used primarily as a reinforcing agent in automobile tyres and many other rubber products. Various manufacturing processes are employed in the production of carbon black, usually obtained as soot from partial combustion of organic compounds. It is the most stable chemical and highly carbonaceous product containing more than 90% carbon with small proportion of hydrogen and oxgen [23]. CB has high specific surface area and electrical conductivity which are highly influenced by the manufacturing procedure. It has been widely used in a number of applications ranging from catalysis to polymers reinforcement (see Fig. 8) [23, 24]. Due to its novel application, constant production of CB from pyrolysis of waste car tyres is highly encouraged. This will not only reduce the market price of CB, but also will provide sustainable solution related to waste tyres management and preventing environmental degradation.

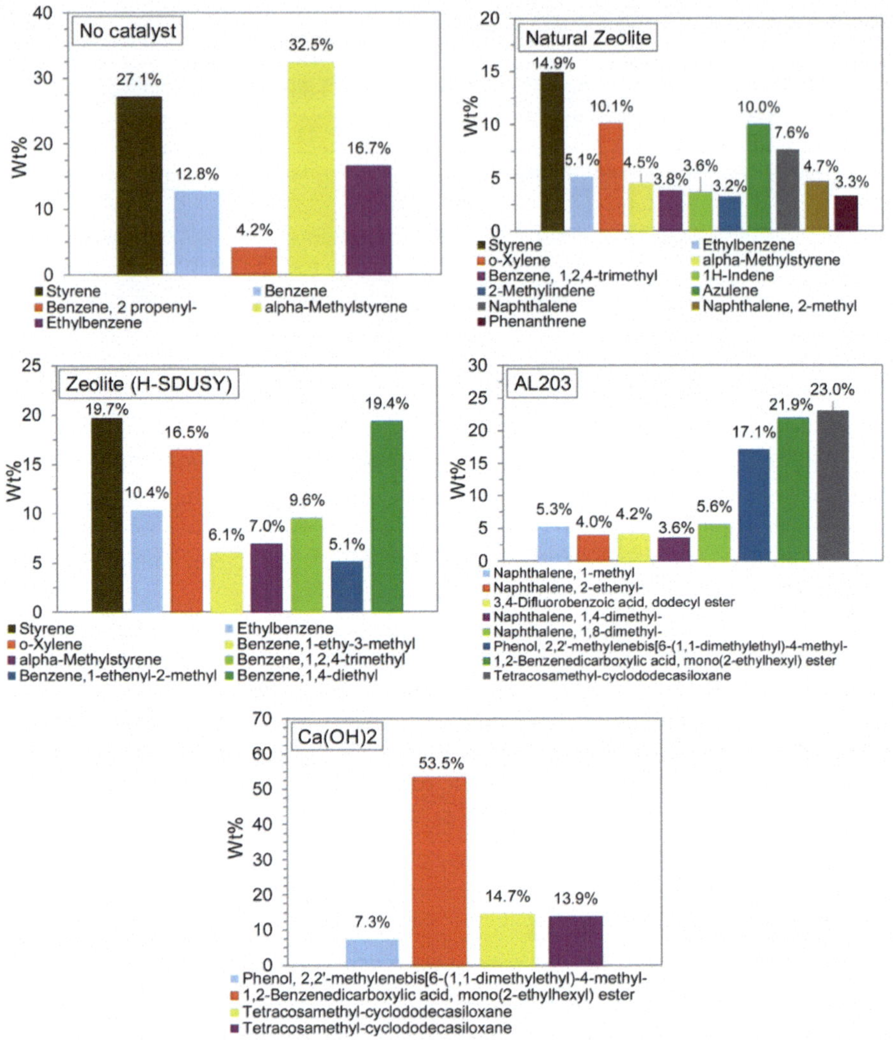


**Fig. 7** GC–MS results of liquid oil produced from catalytic and non-catalytic pyrolysis of tyre waste

### 3.2.1 Properties of CB from Used Tyres

X-ray Diffraction (XRD) patterns of the pyrolytic CB heat-treated at 1100 and 1600 °C from used tyres is shown in Fig. 9a. From the peaks, it can be noted that carbon treated at 1600 °C the peak intensity increased and become rather narrow, indicating increased crystallinity. Li and co-workers [25] studied the R value (peak height divided by background at the position of the peak), the quantity used to quantify the

**Fig. 8** Recycling process of waste tyres into CB and its related goods

proportuiona of graphen layers in carbon materials. The R-value of the 1600 °C heat-treated carbon was 3.46 higher than 2.31 for carbon treated at 1100 °C. According to these R-values of the two heat-treated carbon, higher temperature treating results in a large fraction of parallel graphene layers, thereafter improving properties of the material. Scanning electron microscopy image of a CB is shown in Fig. 9b. The image shows smooth surface due to the deposition of ash and carbonaceous deposits on the surface. It is very important to know the properties of CB before integrating into the device. Graphitization degree and quantity of defects are one of the most important properties required to be known for CB and many other graphitic materials. These properties are determined from the position and width of the D (defect) and G (crystalline graphate) bands together with the intensity ratio of D bands to G bands ($I_D/I_G$) as shown in Fig. 9c–d [26]. Raman spectroscopy technique is normally used to study such properties. The G band is due to the bond stretching modes of all pairs of $sp^2$ atoms in both rings and chains whereas D band is due to the breathing modes of $sp^2$ atoms in the rings. Gnanaraj et al. [27] studied the D band and G band properties of the pyrolytic CB from used car tyres. From their study, the position of the D and G bands were centered at 1368 and 1585 $cm^{-1}$. The Raman spectra of tyre derived carbon samples pyrolyzed at two temperatures 1100 and 1400 °C shows D band positioned at 1363 $cm^{-1}$ for both samples, while G band was centered at 1599 $cm^{-1}$ and 1604 $cm^{-1}$, respectively. However, the width of the peaks in the sample pyrolysed at 1100 °C were broader than those pyrolysed at 1400

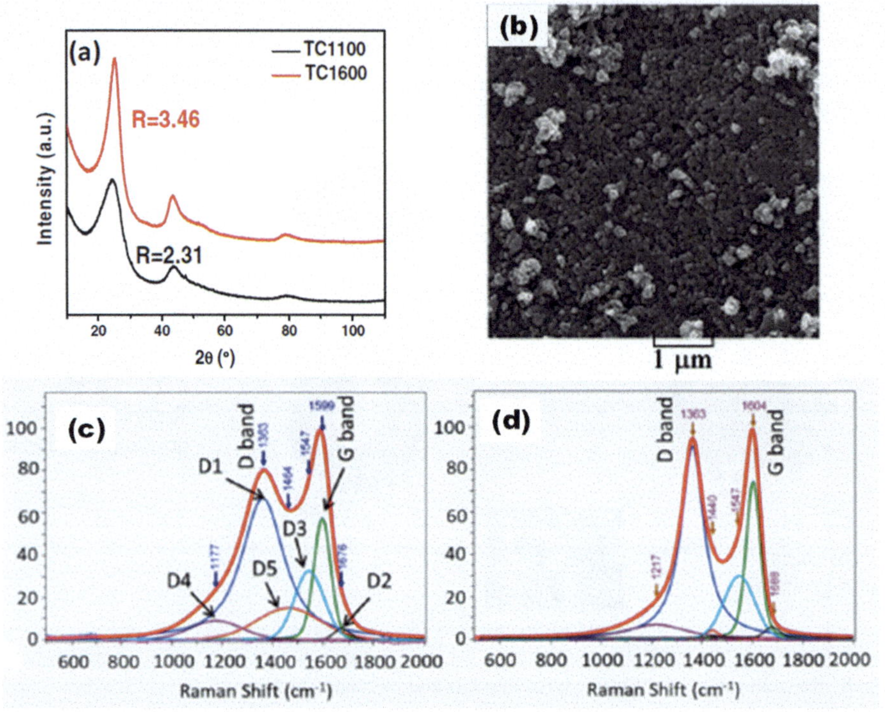


**Fig. 9** **a** XRD patterns waste tyre derived carbon heat treated at 1100 °C (TC1100) and 1600 °C (TC1600), **b** SEM image of pyrolytic carbon black [25], **c**–**d** Raman band of carbon black [26]

°C. The D to G band ratio ($I_D/I_G$) was 0.86 and 0.99 for CB pyrolysed at 1100 and 1400 respectively. The increased $I_D/I_G$ ratio and G band position for the tyre-derived carbon pyrolyzed at higher temperature indicates that more sp2 amorphous carbons turns into nanocrystalline.

### 3.2.2 Uses of BC

Electrochemical Energy Storage

The pyrolysis of used tyres rubber into CB provides many promising applications in electrochemical energy storage systems. The demand for efficient potassium–ion batteries, sodium—ion batteries, lithium–ion batteries to mention, the few, is rapidly increasing in powering electronic devices and vehicles. Figure 10 provides the schematic diagram of the steps involved in the conversion of scrap tyres to a battery grade porous carbon. CB used in almost all batteries found to improve battery performances interms of storage and conductivity. The material can be used as anodes for lithium–ion, lithium–sulfur potassium–ion or sodium–ion batteries. In

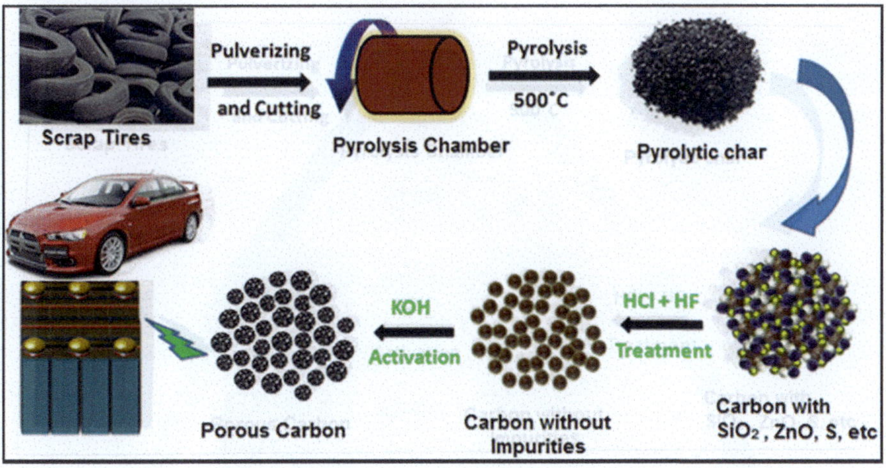


**Fig. 10** Schematic illustration of the processes involved in the pyrolyisis of scrap tyres into battery grade CB [29]

storage batteries, CB manmade material demonstrated a reversible capacity higher than what can be achieved with graphite, a natural carbon [28]. Low cost derived CB demonstrated to be potential anode materials presenting important and interesting application of the used tyres recycling products.

1. Lithium–sulfur batteries have superior properties compared to other batteries. However, they have problems inhibiting battery performances and storage life. Amongst problems, Li–S demonstrated low $Li^+$ ionic conductivity, lithiation of sulfur to $Li_2S$ resulting in a serious volume expansion casing disintegration of electrodes and movements of polysulfides causing internal energy loss which is associated with quick self discharging to mention the few [30]. The above-mentioned problems diminish the batteries performance. As the world is in great demand of clean and cheap energy from solar cells, storage devices are to be good enough and last longer. Among other issues to solve the above-mentioned problems of Li–S batteries, introduction of high conducting material is essential. Carbon related materials are promising ones due to the easy of processing and good electrical conductivity. Apart from being good conductor, they can be also used as barrier to limit sulfur movements (provides confinements and adsorb soluble polysulfides) as depicted in Fig. 11. Pyrolytic CB from used tyres will thus be suitable for batteries as will solve multiple problems including cost, minimize environmental pollution and at the same time having batteries with improved working capability. It has been reported that, pyrolytic CB obtained from scrap tyres is 81.79 wt% carbon reach compared to the 46.03 wt% carbon obtained from high density polyethylene pyrolysis indicating that pyrolytic CB is carbon rich material [31]. Thus CB from scrap tyres is a novel cheap material material in energy storage devices.

**Fig. 11** Schematic diagram representing basic functions of carbon in Li–S batteries (Reference: Carbon materials for Li–S batteries: functional evolution and performance improvement)

### CB in Solar Cells

Silicon based (termed as first generation) and thin films (second generation) solar cells are the ones dominating world's photovoltaic market. However, high manufacturing cost and high energy-consumption processes during the production are the drawbacks silicon solar cells facing. Thin-film solar cells also have shortcomings such as scarcity of raw materials, toxicity of some of the materials involved and high preparation cost. The third-generation solar cells including organic solar cell (OSC), dye-sensitized solar cell (DSSC), quantum dot solar cell (QDSC) and perovskites show great prospect because of their interesting properties such as low-cost, abundant raw materials and less energy consumption during processing. They have the potential of becoming future solar cells as can be synthesized using simple and cheap techniques, implying a great potential to bring down the cost of energy production. Different groups are conducting research on further improving power conversion efficiency (PCE) and stability at the same time minimizing material as well as prepation costs of these third generation solar cells.

Carbon based materials are one of the cheap and readily available material to be used in third generation solar cells as counter electrode (CE) replacing very expensive noble metals previously used. These noble metals are too expensive for mass industrial production and evaporating Au or Ag onto the top of the perovskite solar cell

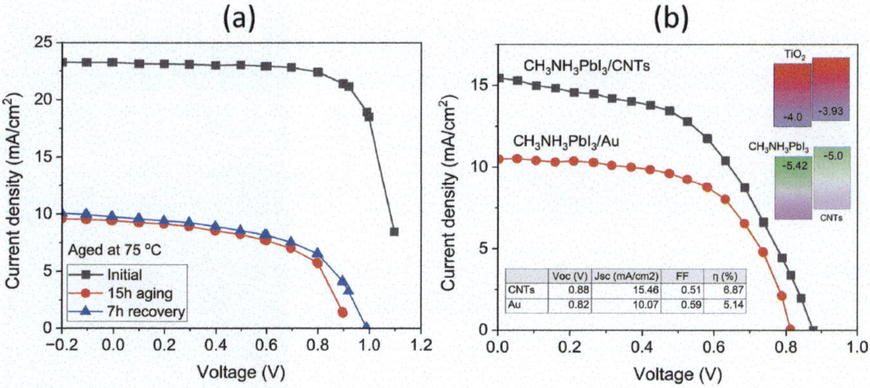


**Fig. 12** PCE perovskite solar cell performance using **a** carbon material as counter electrode and **b** gold contact electrode

is a high energy-consumption process which needs high vacuum and high temperature. Moreover, the vacuum evaporation processes are highly energy consumptive technique and not suitable for low energy consumption processing techniques. Using carbon graphite as CE was first studied by Kay and Graitzel [32] reporting a new type of liquid monolithic DSSC using carbon black/graphite composite CE resulting in the promising PCE of 6.7%.

Recently, Ku et al. [33] reported 6.64% stable PCE for a perovskite solar cell with carbon counter electrode (CCE). This constantly reduces the cost of the overall device. An impressive 12.84 stable PCE was obtained in a novel mixed cation perovskite solar cells employing CCE [34]. Domanski et al. [35] investigated on the stability perovskite solar cells with efficiencies exceeding 20%. They found that exposing perovskite solar cells to a temperature of 70 °C was quite enough to induce gold migration through the hole-transporting layer into the perovskite photoactive material, which successively severely affected the device performance. Figure 12a shows the photovoltaic performance of the device sharply decreased after aging at 75 °C. Figure 12b shows the perovskite solar cell using carbon material as counter electrode and gold. Even though they have been prepared under the same conditions, the device with carbon electrode outperformed the one with gold as clearly seen from the J–V curves.

Due to the outstanding properties CB obtained from used tyres have, they can therefore be used as CCE replacing cormecial CB in perovskite solar cells. Compared with previously used noble metals such as gold (Au) and silver (Ag), CCEs are more cost-effective materials and easier processing techniques. Figure 13a shows schematic illustration of perovoskite solar cell employing graphite/CB as CCE and (b) shows SEM cross section image of the same.

The improved performance of the perovskite solar cells with CCE could be related to the porous nature of carbon which directly affects the particle size and uniformity of perovskites and intrun greatly influence on the photovoltaic perfromance of the

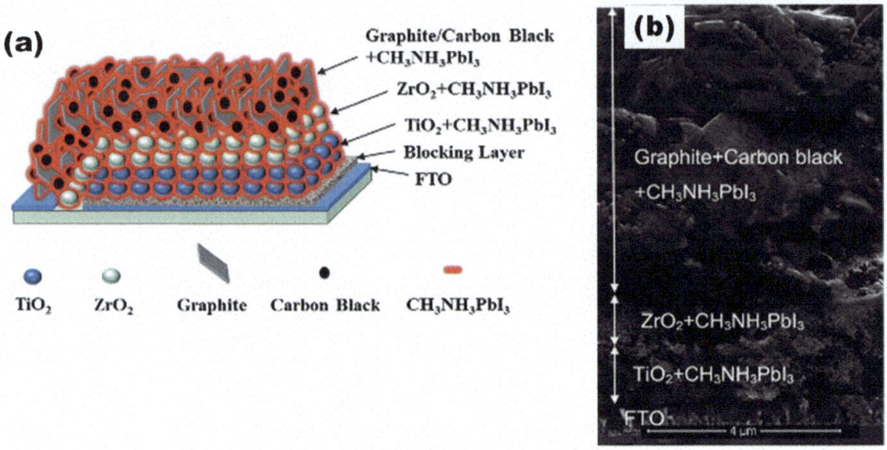


**Fig. 13** Schematic diagram showing perovskite solar cell with graphite/CB as CCE and SEM cross-section image of a $TiO_2/CH_3NH_3PbI_3$ heterojunction perovskite solar cell based on CCE [36]

device. Since carbon materials proved to be very intersting CE, recycling scrap tyres into pyrolytic CB is therefore highly encouraged in the interest of circular economy and constesting the curretn unaffordable market of solar panels.

## 4 Conclusion

The waste tyres are one of the very common and most important hazardous solid wastes disposed everyday all over the world. The management of these waste tyres has become a worldwide environmental challenge. Waste tyres likewise represent a high danger of flames that can be hard to stifle and could have ecological effects because of uncontrolled outflows of conceivably destructive mixes into the air, soil and groundwater. Solving these problems, different secondary and primary uses of waste tyres are undergoing that promote Circular Economy, creation of jobs and minimizing environmental degradation. Recycling of used tyres, a good example of Circular Economy is one of the promising process that will solve all problems associated with scrap tyres disposal. Among the different scrap tyre recycling, pyrolysis is regarded as a promising route. Pyrolytic carbon black, liquid fuel and non-condensed mixture of gases are the product resulted from scrap tyres pyrolysis. Pyrolytic recycling of millions scrap tyres disposed yearly in the country will promote economic growth as large amount of carbon black, pyrolytic oil and syngas will produced. Carbon black is strongly needed as can be incorporated in solar cells as counter electrode and in energy storage devices. Moreover, oils and syngas will also be produced that can be used as fuels and can be further refined into other chemical products. Owing to the great demand of energy in the country, this product will be utilised to capture solar energy and thus genearting clean, renewable energy, obtaining fuels and other useful chemical products and at the same time solving the waste tyres management problems. For that reason, it can be concluded that scrap tyres recycling is not only done to solve environmental problems, but largely promote a Circular Economy and consequently contributing to climate change mitigation.

**Declaration of Conflicting Interests** The author(s) declared no potential conflicts of interest with respect to the research, authorship, and publication of this article.

**Funding** The author(s) received no financial support for the research, authorship, and publication of this article.

## References

1. Wilkinson, A., & Mangalagiu, D. (2012). Learning with futures to realise progress towards sustainability: The WBCSD vision 2050 initiative. *Futures, 44*(4), 372–384.
2. Önenç, S., Brebu, M., Vasile, C., & Yanik, J. (2012). Copyrolysis of scrap tyres with oily wastes. *Journal of Analytical and Applied Pyrolysis, 94*, 184–189.
3. Lopez, G., Olazar, M., Aguado, R., Elordi, G., Amutio, M., Artetxe, M., et al. (2010). Vacuum pyrolysis of waste tyres by continuously feeding into a conical spouted bed reactor. *Industrial & Engineering Chemistry Research, 49*(19), 8990–8997.
4. Nkosi, N., Muzenda, E., Zvimba, J., & Pilusa, J. (2013). *The waste tyre problem in South Africa: An analysis of the REDISA plan.*
5. Hood, Z. D., Adhikari, S. P., Evans, S. F., Wang, H., Li, Y., Naskar, A. K., et al. (2018). Tyre-derived carbon for catalytic preparation of biofuels from feedstocks containing free fatty acids. *Carbon Resources Conversion, 1*(2), 165–173.
6. Leung, D. Y. C., & Wang, C. L. (1999). Kinetic modeling of scrap tyre pyrolysis. *Energy & Fuels, 13*(2), 421–427.
7. Martínez, J. D., Lapuerta, M., García-Contreras, R., Murillo, R., & García, T. (2013). Fuel properties of tyre pyrolysis liquid and its blends with diesel fuel. *Energy & Fuels, 27*(6), 3296–3305.
8. Rathsack, P., Riedewald, F., & Sousa-Gallagher, M. (2015). Analysis of pyrolysis liquid obtained from whole tyre pyrolysis with molten zinc as the heat transfer media using comprehensive gas chromatography mass spectrometry. *Journal of Analytical and Applied Pyrolysis, 116*, 49–57.
9. Sienkiewicz, M., Kucinska-Lipka, J., Janik, H., & Balas, A. (2012). Progress in used tyres management in the European Union: A review. *Waste Management, 32*(10), 1742–1751.
10. Karmacharya, M. S., Gupta, V. K., Tyagi, I., Agarwal, S., & Jha, V. K. (2016). Removal of As (III) and As (V) using rubber tyre derived activated carbon modified with alumina composite. *Journal of Molecular Liquids, 216*, 836–844.
11. Shirule, P. A., & Husain, M. (2015). Reuse of scrap tyre as partial replacement of fine aggregate in concrete and its impact on properties of concrete. *International Journal of Civil and Structural Engineering, 5*(4), 353.
12. Islam, M. R., Joardder, M. U. H., Hasan, S. M., Takai, K., & Haniu, H. (2011). Feasibility study for thermal treatment of solid tyre wastes in Bangladesh by using pyrolysis technology. *Waste Management, 31*(9–10), 2142–2149.
13. Quek, A., & Balasubramanian, R. (2013). Liquefaction of waste tyres by pyrolysis for oil and chemicals—a review. *Journal of Analytical and Applied Pyrolysis, 101*, 1–16.
14. Kar, Y. (2011). Catalytic pyrolysis of car tyre waste using expanded perlite. *Waste Management, 31*(8), 1772–1782.

15. Lah, B., Klinar, D., & Likozar, B. (2013). Pyrolysis of natural, butadiene, styrene–butadiene rubber and tyre components: Modelling kinetics and transport phenomena at different heating rates and formulations. *Chemical Engineering Science, 87*, 1–13.
16. Bakker, M. M., Govers, G., van Doorn, A., Quetier, F., Chouvardas, D., & Rounsevell, M. (2008). The response of soil erosion and sediment export to land-use change in four areas of Europe: The importance of landscape pattern. *Geomorphology, 98*(3–4), 213–226.
17. Laboy-Nieves, E. (2014). Energy recovery from scrap tyres: A sustainable option for small islands like Puerto Rico. *Sustainability, 6*(5), 3105–3121.
18. Samolada, M. C., & Zabaniotou, A. A. (2012). Potential application of pyrolysis for the effective valorisation of the end of life tyres in Greece. *Environmental Development, 4*, 73–87.
19. Su, Y., & Deng, W. (2010, November). A thermogravimetric study of waste tyre powder. In *2010 International Conference on E-Product E-Service and E-Entertainment* (pp. 1–4). IEEE.
20. Paasikallio, V., Lindfors, C., Kuoppala, E., Solantausta, Y., Oasmaa, A., Lehto, J., et al. (2014). Product quality and catalyst deactivation in a four day catalytic fast pyrolysis production run. *Green Chemistry, 16*(7), 3549–3559.
21. de Marco Rodriguez, I., Laresgoiti, M. F., Cabrero, M. A., Torres, A., Chomon, M. J., & Caballero, B. (2001). Pyrolysis of scrap tyres. *Fuel Processing Technology, 72*(1), 9–22.
22. Miandad, R., Barakat, M. A., Rehan, M., Aburiazaiza, A. S., Gardy, J., & Nizami, A. S. (2018). Effect of advanced catalysts on tyre waste pyrolysis oil. *Process Safety and Environmental Protection, 116*, 542–552.
23. Borah, D., Satokawa, S., Kato, S., & Kojima, T. (2008). Characterization of chemically modified carbon black for sorption application. *Applied Surface Science, 254*(10), 3049–3056.
24. Ciobanu, M., Lepadatu, A. M., & Asaftei, S. (2016). Chemical and electrochemical studies of carbon black surface by treatment with ozone and nitrogen oxide. *Materials Today: Proceedings, 3*, S252–S257.
25. Li, Y., Adams, R. A., Arora, A., Pol, V. G., Levine, A. M., Lee, R. J., et al. (2017). Sustainable potassium-ion battery anodes derived from waste-tyre rubber. *Journal of The Electrochemical Society, 164*(6), A1234–A1238.
26. Maslova, O. A., Ammar, M. R., Guimbretière, G., Rouzaud, J. N., & Simon, P. (2012). Determination of crystallite size in polished graphitized carbon by Raman spectroscopy. *Physical Review B, 86*(13), 134205.
27. Gnanaraj, J., Lee, R., Levine, A., Wistrom, J., Wistrom, S., Li, Y., et al. (2018). Sustainable waste tyre derived carbon material as a potential anode for lithium-ion batteries. *Sustainability, 10*(8), 2840.
28. Retrieved January, 2019, from https://www.machinedesign.com/batteriespower-supplies/recycling-used-tyres-battery-electrodes.
29. Kumar, R., & Sharma, A. (2018). Morphologically tailored activated carbon derived from waste tyres as high-performance anode for Li-ion battery. *Journal of Applied Electrochemistry, 48*(1), 1–13.
30. Liang, J., Sun, Z. H., Li, F., & Cheng, H. M. (2016). Carbon materials for Li–S batteries: Functional evolution and performance improvement. *Energy Storage Materials, 2*, 76–106.
31. Zhang, X., Li, H., Cao, Q., Jin, L. E., & Wang, F. (2018). Upgrading pyrolytic residue from waste tyres to commercial carbon black. *Waste Management & Research, 36*(5), 436–444.
32. Kay, A., & Grätzel, M. (1996). Low cost photovoltaic modules based on dye sensitized nanocrystalline titanium dioxide and carbon powder. *Solar Energy Materials and Solar Cells, 44*(1), 99–117.
33. Ku, Z., Rong, Y., Xu, M., Liu, T., & Han, H. (2013). Full printable processed mesoscopic $CH_3NH_3PbI_3/TiO_2$ heterojunction solar cells with carbon counter electrode. *Scientific Reports, 3*, 3132.
34. Mei, A., Li, X., Liu, L., Ku, Z., Liu, T., Rong, Y., et al. (2014). A hole-conductor–free, fully printable mesoscopic perovskite solar cell with high stability. *Science, 345*(6194), 295–298.
35. Domanski, K., Correa-Baena, J. P., Mine, N., Nazeeruddin, M. K., Abate, A., Saliba, M., et al. (2016). Not all that glitters is gold: Metal-migration-induced degradation in perovskite solar cells. *ACS Nano, 10*(6), 6306–6314.

36. Xu, X., Liu, Z., Zuo, Z., Zhang, M., Zhao, Z., Shen, Y., et al. (2015). Hole selective NiO contact for efficient perovskite solar cells with carbon electrode. *Nano Letters, 15*(4), 2402–2408.

# Low Carbon Energy Systems

**Nhlapo P. Lefu and Motaung Tshwafo**

**Abstract** Low carbon energy systems describe a widely used set of approaches used as an alternative means to generate energy with a view to curbing the emission of greenhouse gas (GHG). Amongst these are hydro, solar, wind and biomass. The fossil fuels have been in use to generate electricity since at least mid-1920s. The industry is one of the main contributing direct factors towards the observed changes in global mean temperature, rainfall patterns and storms already globally. And this led to the drafting of the landmark Paris agreement which seeks to strengthen the international community response to climate change. This chapter presents a global heat analysis and contribution by NASA. More information about African continent in terms of the precipitation pattern, the rise in sea level, and the impact of climate change on food security and health are thoroughly discussed.

**Keywords** Low carbon energy system · Global heat · Legislatures · Renewable energy sources

## 1 Introduction

Apart from energy being essential in meeting basic human needs it has been the fundamental component of economic growth and socio-economic development since the industrial revolution that started in Europe during the 18th century [1, 2]. World energy combustion of fossils fuel such as coal, oil and gas (which emit greenhouse gas (GHG)) to generate energy is not a new thing, the method has been used since the middle of the 20th century [2]. The energy industry is arguably the largest contributor to global GHGs emissions as the current electricity generating infrastructure is still overwhelmingly dependent on carbon intensive primary fuels such as coal and natural

N. P. Lefu
Sasol Operations, P.O. Box 1, Sasolburg, Gauteng, South Africa

M. Tshwafo (✉)
Department of Chemistry, Sefako Makgatho Health Sciences University, P.O. Box 94, Medunsa 0204, South Africa
e-mail: tshwafo.motaung@smu.ac.za; motaungte@live.com



M. Tshwafo Elias (ed.), *Waste to Value Innovations*,
https://doi.org/10.1007/978-981-96-2782-0_4

gas. The other sources of GHGs being the anthropogenic components. The rising GHGs viz. (i) carbon dioxide ($CO_2$) and (ii) non-$CO_2$ in the atmosphere led to increased global mean surface temperature. The non-$CO_2$ include, inter alia, the following:

- methane ($CH_4$),
- nitrogenous oxide ($N_2O$),
- and fluorinated gases (three major groups being the hydroflourocarbons, perflourocarbons, and sulfur hexafluoride).

The increased global mean surface temperature has a detrimental effect on climate systems, on the natural environment, and upon human society leading to a phenomenon called the climate change [3]. The effects of growing threat of dangerous climate change which include the melting of icecaps (in the polar areas), increased changeability in temperature, rainfall and storms (which encompasses the increasing intensity and frequency of climate extremes) are already being felt across the world [4, 5].

The prospect of climate change demands a significant reduction in the use of fossil-related $CO_2$ emissions energy. With that in mind, Parthan et al. [6], outlined that the primary energy demand projections from the International Energy Agency predicted the global energy-related $CO_2$ emissions would increase from 28.8 Gigatonnes (Gton) of $CO_2$ in 2007 to 40.2 Gton of $CO_2$ equivalent in 2030. In broader terms, such an increase would result in a long-term concentration of above 1000 ppm $CO_2$ equivalent and a global average temperature rise of 6 °C. In response to that, the landmark Paris Agreement was drafted in 2015, from which the international community agreed to strengthen its response to climate change. The parties agreed to keep rise in global mean temperature well below 2 °C above pre-industrial levels and will pursue efforts to limit the increase to 1.5 °C [7].

Subsequent to the adoption of the Paris agreement in November 2016, Bello and Ojoyi [8], also estimated the $CO_2$ emissions of global reserves of fossil fuels to Ca. 11 000 Gton. They stated that the set limit of 2 °C is achievable by using the fossils fuel in the region of less than or between 870 to 1240 Gton of these reserves (of 11 000 Gton) between the 2011 and 2050. Similarly, Shen et al. [9] also indicated that the targets set by the Paris agreement were not farfetched provided that the $CO_2$ emissions is controlled under 1100 Gton, which implies that Ca. 85%, 50% and 34% of global coal, gas and oil reserves respectively should not be used before 2050. According to Blanco et al. [2], both the global surface temperature and global mean sea level had already increased by 0.9 °C and 0.2 m respectively compared to pre-industrial times. These researchers further indicated that the temperature can be limited to 2 °C by 2100; if and only if; the cumulative emissions over the period of 2012–2100 stay within 1000 Gton $CO_2$. This would imply the accelerated need to transition to renewable energy sources.

## 2 Global Heat

According to NASA's Goddard, Institute for Space Studies (GISS) the world is getting warmer regardless of the sources [10]. Nonetheless the Industrial Revolution was identified as a main role player influencing the thermometer. In a short space of time since 1880 the average global temperature on earth has increased by about 0.8 °C (1.4 °F) at a rate of roughly 0.15–0.20 °C per decade. The average temperature depends on how much energy the planet receives from the sun and how much it radiates back into space, whereas the amount of energy radiated by the earth depends significantly on the chemical composition of the atmosphere, mainly the amount of heat-trapping greenhouse gases. For a one-degree global change there should be a vast amount of heat to warm all the oceans, atmosphere, and land by that much.

Figure 1 shows yearly temperature anomalies from 1880 to 2014 as recorded by the Japan Meteorological Agency, and the Met Office Hadley Centre (United Kingdom), NASA Goddard Institute for Space Studies and NOAA National Climate Data Centre. Generally, all four records show peaks and valleys in sync with each other. All show rapid warming in the past few decades, and all show the last decade as the warmest. The levelling off between the 1940 and 1970s was attributed to the natural variability and possibly by cooling effects of aerosols generated by the rapid economic growth after World War II. An aerosol cooling is more immediate, while greenhouse gases accumulate slowly and take much longer to leave the atmosphere. The strong warming trend of the past three decades apparently reflects a shift from comparable aerosol and greenhouse gas effects to a prevalence of greenhouse gases.

Further note that, it is on NASA's records that since the year 2000, land temperature changes are 50% greater in the United States than ocean temperature changes; two to three times greater in Eurasia; and three to four times greater in the Arctic and the Antarctic Peninsula. Warming of the ocean surface has been largest over the Arctic Ocean, second largest over the Indian and Western Pacific Oceans, and third largest over most of the Atlantic Ocean.

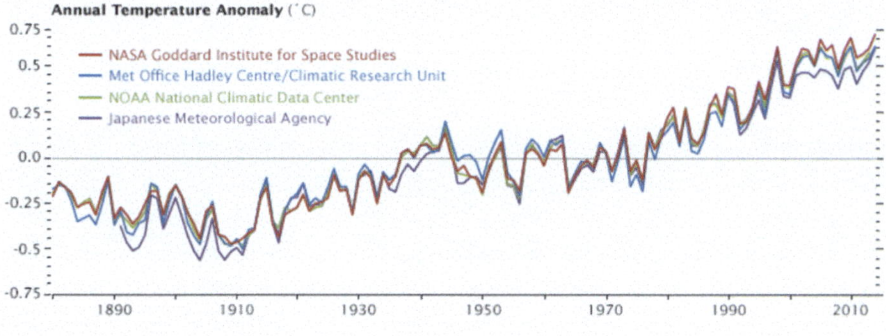


**Fig. 1** Annual temperature anomalies

## 3 African Heats

According to Serdeczny et al. [11], Africa has been identified as one of the most vulnerable to the impacts of climate change. Thus, there are developing projections in Africa, as well, which obviously contribute on putting the world on a path toward undesirable warming. For example, sub-Saharan Africa has experienced more frequent and more intense climate extremes over the past decades [12]. However, changes are not constant across the region. East Africa is apparently at high risk of flooding and concurrent health impacts and infrastructure damages [13]. West Africa is projected to experience risks for food security and negative ramifications for human health and poor marine production. While republic of South Africa could experience the strongest decrease in precipitation with concurrent risks of drought. The potential interactions and amplifying effects across sectors are not yet reflected in the science [11]. Integrating impacts across sectors and population dynamics remains a major challenge for science. Climatic changes and impacts across sectors at different levels of warming are summarized in Fig. 2.

Of course there are boundaries within the laws of chemistry and physics to curb global temperature rise to 1.5 °C, but doing so would require exceptional changes in agriculture, energy, industry, buildings, transport, and infrastructure in general. This of course would also change the landscape of regional and continental economy. United Nations Framework Convention on Climate Change finalised the rules for implementation of strategies to encounter the global temperature increase. This is

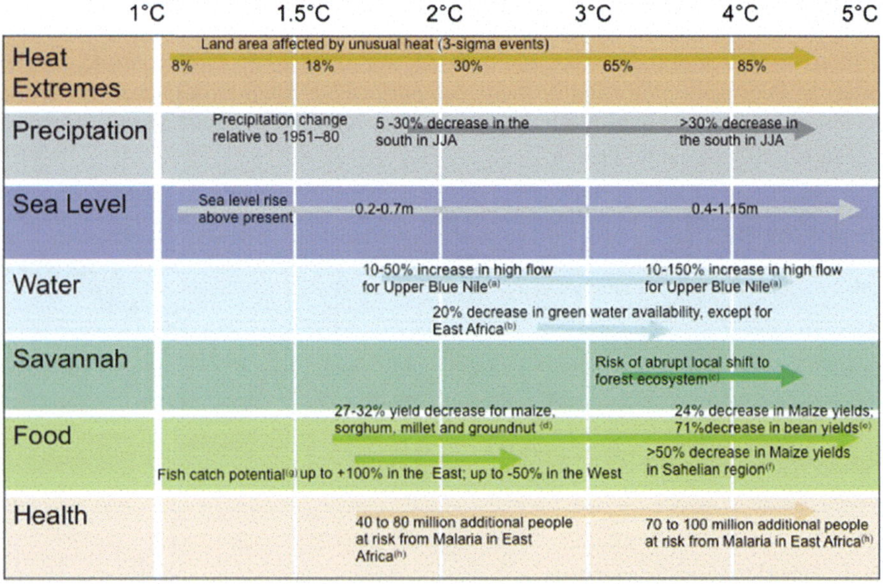


**Fig. 2** Climatic changes and impacts across sectors at different levels of warming

after climate change has already exacerbated existing problems, including conflicts, migration and shrinking natural resources such as water and land. For instance, the shrinking of lakes at the boarders lead to economic marginalization and provide a breeding ground for recruitment by terrorist groups as social values and moral authority evaporate. This may be the part of explanations why Boko Haram (So-called Nigerian terrorist) existed and have operated for some time. Sahel, Limpopo basin and areas of the Zambezi basin in Zambia depend on rain-fed agriculture and are regularly hit by droughts and floods, with enormous consequences food security. The clear evidence of water scarcity has shown in South Africa since 2000. In 2019 people were delivered water with containers in both suburbs and villages. There would not be a silver human being if the global warming is not effectively monitored and controlled. According to the IPCC, projections show that the western Sahel region will experience the strongest drying, with a significant increase in the maximum length of dry spells. The IPCC expects Central Africa to see a decrease in the length of wet spells and a slight increase in heavy rainfall [12].

### *3.1 Renewable Energy Sources (Research and Developments)*

Seeing that the major component of GHGs is $CO_2$, the international communities and national governments are gearing up to reduce carbon emissions. Countries tailored the policies that would drive the enhancement of renewable energy development and encouragement of technological innovations [14]. It is reported elsewhere Boyu [15] that the renewable energy sources (also known as carbon neutral technologies) are being referred to as energy sources that can be generated and sustainably used in nature, including non-fossil energy such as wind energy, solar energy, waterpower, biomass energy, geothermal and ocean energy, etc. With that in mind, Mathieu [16] and Jenniches [17] outlined that many governments have developed supporting mechanisms such as feed-in tariffs, renewable portfolio standards and tax policies etc. for development of renewable energy generation, in tandem, with implementing energy use efficiency for saving energy.

Many countries have embarked on installing facilities that use renewable energy sources for power generation. The drive to transition to renewable energy technologies is said to be enthused by four factors viz. energy security, economic impacts, $CO_2$ reduction [2, 17] and encouraging economic development [18]. According to Chinese lawyer [19], majority of countries regard the exploitation of renewable energy sources as an evolution for developing circular economic strategy which would lead and standardize the development of renewable energy sources by means of the law. Different studies have outlined the strategies and policy set by different countries in pursuit of the realisation of targets being stipulated in the Paris Agreement. van der Zwaan and Longa [20] explored the attainability of the large-scale renewable energy technology deployment against Kenya's GHG reduction ambitions by paying a specific focus on the energy demand and power sector. The study was conducted as an attempt to answer the simple question being asked "Do Kenya's climate change mitigation ambitions necessitate large-scale renewable energy deployment and dedicated low-carbon energy policy?". Their study was necessitated by The Kenyan Environmental Management Authority who projected that the geothermal energy alone could yield a GHG reduction of Ca. 14 $MtCO_2e$ in 2030 compared to 1.4 and 1.0 $MtCO_2e$ respectively for wind and solar, in that year. The authors supported the practicability of Kenyan climate management goals, provided that adequate investments in renewable and other low-carbon energy technologies were timely available.

In their study, Sharvini et al. [21] evaluated the challenges and opportunities of energy consumption trends and their linkages with renewable energy policies in East (China and Japan) and Southeast (Malaysia and Indonesia) countries. The countries being varied greatly in terms of characteristics of resources, topography and gross domestic product (GDP) per capita and that made their study to be more interesting. Of the four countries, both Malaysia and Indonesia were said to be richly blessed with natural resources and are top producers of palm oil in which palm oil residues can be converted to energy. Whilst China and Japan were the most developed in terms of GDP of the four countries. The authors reported that though these countries have progressed towards the establishment of rigid energy policies, there were still evident shortfalls on the utilisation of renewables in their energy mixes. Further note that, according to the feedback given by the aforesaid authors, the government and other authorities were said to have made efforts to make subsidies and tax incentives available to encourage the development of renewable energy.

In the recent past, Aliyu et al. [22] reviewed the energy development in Africa focusing on the three leading nations viz. South Africa, Egypt and Nigeria. Succinctly, the three countries combined were said to have about half of the total Africa's primary energy. The authors delineated the significance of the major renewable energy sources (hydro, solar, wind and biomass), energy efficiency of the three countries, energy policy, overview of conventional energy, strategies in place towards using of energy efficiency in Africa, renewable energy road map in Africa, needs for grid extension, energy storage system and seasonal variation in Africa. In Egypt the renewable energy sources contributed 14,855 GW h electricity production in 2012. Of the 14,855 GW h, the 13,358, 1260 and 237 GW h were respectively produced by hydropower, wind energy and solar photovoltaic. According to the information being reported by Tan et al., the three countries have not fully exploited their potential as far as renewable energy is concerned. This is despite Egypt having established their renewable energy authority in the 80s, and South Africa's renewable energy policy being ratified in 2003. The Nigerian government approved the Renewable Energy Master Plan (REMP) shortly thereafter (in 2005). The failure of the three countries to fully exploit their potential is as a result of limited policy interest, investment level and poor infrastructural support necessary to harness renewable energy. Of the three countries, South Africa and Egypt were found to contribute significantly towards renewable energy with target of 42 and 22% renewable energy contribution in electricity generation by the year 2020 and 2030. However, the authors concluded that of the said countries, South Africa has the best model for tacking its energy challenges via the utilization of wind and solar energy relative to its counterparts.

In recent years, there has been an urgent need to diminish the dependency of South and Central America power sectors on a changing hydrological profile. Breyer and coworkers [23] used the energy system optimisation model to evaluate the possibilities of having hydro, wind and solar power as a base for a 100% renewable energy supply for South and Central America in 2030. The authors stated that the hydropower dominated based on the cost minimised design of the energy mix (in terms of TWh of electricity production) study. However, solar photo voltaic technology emerged as the main energy supply in most of the sub-regions (in terms of GW of installed capacities) of South and Central America.

## 4 Conclusion

The developed countries seem to have made a momentous stride for replacing the conventional fossil fuels power generation systems with the low carbon systems. Other countries are still struggling to effectively enforce and/or transition towards low energy systems either because of lack of financial aid or the country's economies are heavily reliant on the fossil fuels production and export. Countries whose economies depend on conventional fossil fuels, viz. coal, gas and oil are slow to accelerate change towards the low carbon energy systems as this would have a detrimental impact on their economic development.

The growth in electricity demand and population growth, inter alia, form an integral part of the limitations in some developing countries thereby lowering the pace of transitioning towards the low carbon power stations. Suffice is to say, the global struggle towards the utilization of low energy systems and subsequent limit to the rising of global temperature to well below 2 °C needs consented efforts from a diverse group of individuals for the energy and climate communities, investors, energy companies, equipment suppliers, energy intensive companies, academics, non-profit organizations and so forth across the world.

## References

1. Mondal, A. H., Ringler, R. M. C., & Pradesha, A. (2018). The Philippines energy future and low-carbon development stratergies. *Energy, 147*, 142–154.
2. Blanco, H., Nijs, W., Ruf, J., & Faaij, A. (2018). Potential for hydrogen and power-to-liquid in a low-carbon EU energy system using cost optimization. *Applied Energy, 232*, 617–639.
3. Nakata, T., Silva, D., & Rodionov, M. (2011). Application of energy system for designing a low-carbon society. *Progress in Energy and Combustion Science, 37*, 462–502.
4. Ivan, N., Gordon, M., Mbogo, A. S., & Zhou, P. (2019). Regional developments in energy systems, economics and climate, Risø energy report 7. DTU orbit, January 17.
5. Jinfang, L., Shunsheng, & Qin, S. (2016). On low-carbon technology. *Low Carbon Economy, 2016*(7), 107–115.

6. Parthan B., Osterkorn M., Matthew, Kennedy M., Hoskyns J., Bazilian M, et al. (2010). Lessons for low-carbon energy transition: Experience from the renewable energy and energy efficiency partnership (REEEP). Energy for sustainable Development
7. Adoption of the Paris Agreement; United Nation (FCCC/CP/2015/L.9/Rev.1).
8. Bello, O., & Ojoyi, M. (2017). Low-carbon energy transition in Africa: Insights from West Africa: Policy Briefing 164. South African Institute of International Affairs.
9. Shen, W., Qiu. J., & Dong, Z. (2018). Electricity network planning targeting low-carbon energy transition. *Global Energy Interconnection, 1*(4), 487–499.
10. Retrieved November 27, 2019, from https://data.giss.nasa.gov/gistemp/.
11. Serdeczny, O., Adams, S., Baarsch, F., Coumou, D., Robinson, A., Hare, W., Schaeffer, M., Perrette, M., & Reinhardt, J. (2019). Climate change impacts in Sub-Saharan Africa: From physical changes to social repercussions. *Regional Environmental Change, 15*(8), 1–15.
12. Retrieved November 27, 2019, from https://www.un.org/africarenewal/magazine/december-2018-march-2019/global-warming-severe-consequences-africa.
13. Retrieved November 27, 2019, from https://igad.int/divisions/agriculture-and-environment/2283-flood-risks-to-continue-across-east-africa.
14. Abolhosseini, S., Heshmati, A., Altmann, J. (2014). A review of renewable energy supply and energy efficiency technologies. Iza DP No. 8145.
15. Boyu, Z. (2011). Further discussion on exploitation of renewable energy sources and its legal regulations. *Energy Procedia, 5*, 2114–2119.
16. Mathieu, C. France: Reducing nuclear dominance and promoting a low-carbon energy system. France (IASS Study_45).
17. Jenniches, S. (2018). Assessing the regional economic impacts of renewable energy sources–a literature review. *Renewable and Sustainable Energy Reviews*, (93), 35–51.
18. Konadu, D. D., Mourão, Z. S., Allwood, J. M., Richards, K. S., Kopec, G. M., McMahon R. A., et al. (2015). Not all low-carbon energy pathways are environmentally "no-regrets" options. *Global Environment Change*, (35): 379–390.
19. Benedek, J., Sebestyén, T. -H., & Bartók, B. (2018). Evaluation of renewable energy sources in peripheral areas and renewable energy-based rural development. *Renewable and Sustainable Energy Reviews*, (90), 516–535.
20. van der Zwaan, B., Longa, F. D. (2017). Do Kenya's climate change mitigation ambitions necessitate large-scale renewable energy deployment and dedicated low-carbon energy policy? *Renewable Energy*, (113), 1559–1568.
21. Sharvini, S. R., Noor, Z. Z., Chong, C. S., Stringer, L. C., & Yusuf, R. O. (2018). Energy consumption trends and their linkages with renewable energy policies in East and Southeast Asian countries: Challenges and opportunities. *Sustainable Environment Research, 28*, 257–266.
22. Aliyu, A. K., Modu, B., Tan, C. W. (2017). A review of renewable energy development in Africa: A focus in South Africa, Egypt and Nigeria. *Renewable and Sustainable Energy Reviews.*
23. Barsova, L. S., Bogdanov, D., & Vainikka, P. (2017). Hydro, wind and solar power as a base for 100 % renewable energy supply for South and Central America. *PLoS ONE, 12*(3), 1–28.

# Fabrication of Electrodes for Lithium-Ion Battery Using Waste Materials

**Luyanda Lunga Noto**

**Abstract** We are constantly generating products to advance human life, and unfortunately, most of these products end up as waste that is polluting our environment. We also rely on fossil fuels to drive our daily lives and development. The consequence of the fossil fuels has been the generation of a large quantity of carbon dioxide waste, which has been deposited into the atmosphere. This deposition is the primary cause of global warming, which is continuously deteriorating our environment through droughts that limit our farming capacity, floods and acid rain that destroy our infrastructure. We have also rendered a portion of our land infertile and therefore reduced the surface area of farmable land, through the large amounts of waste that we are burying underground. In this study I am discussing what is waste, different ways in which our waste pollutes our environment, methods of regulating waste, and more specifically how to convert some forms of waste into active electrodes of a lithium-ion battery.

**Keywords** Lithium ion battery · Waste management · Battery from recycled waste · Waste-to-economic-development

## 1 Understanding Waste

Waste can generally be described as any substance that is no longer useable, and which is no longer needed. It is commonly derived from our households, from the industrial activities and from hospitals, just to mention a few generators of waste [1]. We normally limit it to spoiled food, body excretion or any other thing that we throw away in our homes [2]. However, waste is more than that and it can be in a gas, liquid or solid form, as discussed below.

i. Water pollution

L. L. Noto (✉)
Department of Physics, University of South Africa, Private Bag X90, Florida 1710, South Africa
e-mail: notoll@unisa.ac.za

 
M. Tshwafo Elias (ed.), *Waste to Value Innovations*,
https://doi.org/10.1007/978-981-96-2782-0_5

**Fig. 1** Sewage water disposed into the open land [3]

Dirty water is commonly generated in our homes, which we dispose as waste, and due to the poor sewage system in most townships, this water ends up in rivers, as pollutant [3]. A greater amount of this type of waste is generated by the hospitals, industries and the mining sectors (Fig. 1). The textiles industries get rid of water with different dyes, and mines dispose a lot of water after cooling the heated ore [4].

ii. Solid wastes

This type of waste includes disposed toilet paper, news-papers, magazines and plastic grocery bags from our households (Fig. 2), just to mention a few [5]. The industries produce a lot of boxes and plastics that are used to wrap cargo, for shipment from the industry to the retail or end users, to preserve the product. Unfortunately, though the primary use of the plastics and boxes is aimed at achieving something good, they somehow find a way to the streets, parks and rivers as waste [6].

Households generate waste in the form of empty food cans, bottles of detergents, broken chairs and glasses, just to mention a few [5]. The beverage industry is by far producing a lot of glass carrying juice, fizzy and alcohol type of drinks. These bottles end up as waste that is polluting our environment (Fig. 3) [7].

Another common household waste is from organic food, which include vegetable skins, spoiled vegetable and fruits (Fig. 4) [5]. The industry does produce a lot of organic waste too, through food processing, manufacturing of paper and pharmaceuticals, and more other organic products. Though organic waste is degradable overtime, but it does affect children who attempt to eat the spoiled fruits [8].

**Fig. 2** The common plastic, box and paper waste material [6]

**Fig. 3** Broken bottle waste polluting the environment [7]

**Fig. 4** Organic waste made up of different fruits [9]

iii. Hazardous waste

The hazardous waste is mostly undesirable because it threatens the well-being of human and animals. It includes gases, liquids or solids that corrode the environment and flammable, toxic and corrosive substances (Fig. 5). This type of waste is mostly generated by industries, hospitals and mines, through an emission of toxic gases [1]. Some of the hazardous substances, are available in our households, which when not used properly can harm us and the animals around us [5]. The carbon emission from the cars and the industry, forms part of the human waste, which is devasting our atmosphere [3, 8].

## 2 Side-Effect of Waste Material

Improper management of waste material can lead to several side effects to the environment and its habitants. Children can be badly injured if they were to play in a field filled with a lot of broken bottles, and they may suffer from serious illnesses if they were to drink water contaminated by sewage. There are several reported incidences where serious damage to the environment occurred due to waste, as discussed below.

**Fig. 5** Mixture of different hazardous types of waste [10]

i. Water contamination effects

Water is a solvent and it tends to dissolve all chemical materials soluble in it, which may include several toxic and corrosive chemicals. Such contaminated water may find its way to dams, rivers and the ocean (Fig. 6) through acid rains [8] or uncontrolled industrial drainage systems [3]. It is common in most townships to find children playing in dams and people drinking from rivers in villages. This immediately subjects them to all the toxic chemicals and bacteria from organic waste present inside the water which they drink. In South Africa we usually have outbreaks of Cholera, which have now claimed several lives due to drinking contaminated water [11].

There are several landfills across South Africa, where the waste material is continually burnt. Unfortunately, due to the growth of towns and cities, our houses have moved closer to the landfills, which were initially far. And the lack of relocating such sites, results into people coming into contact with the released gases when the waste is being burnt. The landfill gases have been reported to cause cancer, respiratory and eye problems. The toxic gases cause skin irritation, and continuous scratching leads to damaged skin tissues [12].

ii. Marine pollution

Ocean pollution is one of the problems set up by human, which threatens the marine environment (Fig. 6). Though the shells of dead ocean creatures do pollute the shore [13], the effect is not as catastrophic as that of the pollution caused human activities. The human pollution of the ocean disturbs the normal living conditions of the marine life and limits the population of certain marine species [14].

The human activities continue to pollute the marine environment in many ways, which include: The inflow of rivers infected by acidic mine drainage and municipality sewage, unregulated municipality landfills that are built near the marine, which result in debris flying into the ocean, unregulated littering of plastics and oil spills from the ships transporting crude oil from country to another [14].

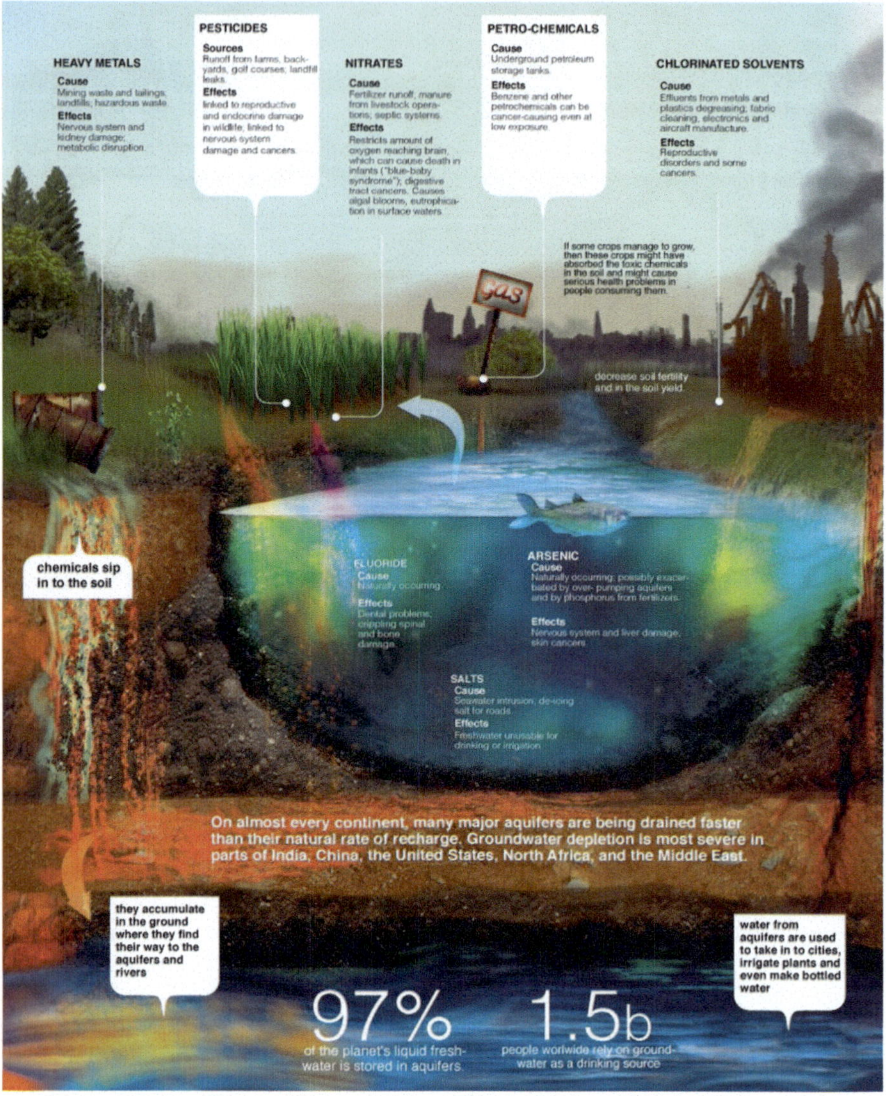


**Fig. 6** An image showing various forms of environmental pollution [15]

iii. Soil contamination

The soil is often degraded off its natural minerals by the unregulated waste that is disposed randomly onto the ground, such that it negatively affects its natural ecological status [16]. Ideally, all plastics, bottles and other industrial chemical waste need to be regulated. However, it is not always the case. The plastic and bottle waste that is buried underground, and the industrial acidic drainage sinking into the ground (Fig. 6), render the soil in fertile. The common chemicals that sink into the soil include pesticides in the agricultural sector, buried radioactive waste, petroleum oil and many other toxic heavy metals from the industry and mines [17].

iv. Air pollution

Without a doubt, we are all feeling the side effects of global warming, through several heat waves that cut across South Africa, the prolonged drought in the Free State Province, the dry rivers across the country and the rising sea levels due the melting ice at the poles, just to mention a few. All these are attributed to global warming, which is a result of greenhouse gases emitted into the atmosphere due human activities here on earth [18].

The greenhouse gases include Water vapor ($H_2O$), Carbon dioxide ($CO_2$), Methane ($CH_4$), Nitrous oxide ($N_2O$), Ozone ($O_3$), Chlorofluorocarbons (CFCs), and Hydrofluorocarbons. Gases like carbon dioxide and methane have the capacity to trap heat, and their increment in concentration results in more heat being trapped. On earth we continue to generate carbon dioxide and methane emission from fossil fuels (Fig. 6), and we cut down forests that act in reducing the carbon dioxide in the atmosphere. This increases the quantity of Carbon dioxide and Methane in the atmosphere, and therefore increasing global warming [18].

## 3 Waste Management Systems

To minimize the side effects of waste in our environment, necessary procedures need to be taken to ensure that every household, hospital, research institute, mine or industry, fully takes responsibility of its waste. The government should make knowledge about waste management mandatory to all citizens. Each company should have well trained waste management representatives and each household should have atleast one member who has undergone a crash course in waste management.

Even though the waste management procedures are not familiar to everyone, there is an existing process that has been adopted to manage waste [19], even though it's not fully adhered to. In South Africa, The National Environmental Management; Waste Act No. 59 of 2008 [1], was established to manage waste according to the global waste hierarchy. The process is not as effective as it is in first world countries, where ordinary citizens are aware of waste management and recycling of the household waste.

If well implemented, the global waste management system (Fig. 7) can see our environment waste free. The model first insists on preventing waste creation [1, 19]. It would be beneficial in managing waste if an item for sale was rated according to its reusability and recyclability, to guide the consumer into purchasing only reusable and recyclable items. This would assist in preventing and minimizing waste formation. Waste prevention also entails using less materials to fabricate the product and encourages effective use of the product so that it does not end up as waste [20].

The system encourages that in cases where it is unavoidable to buy an item that may end up as waste material, then items with longer lifetime and fabricated with less

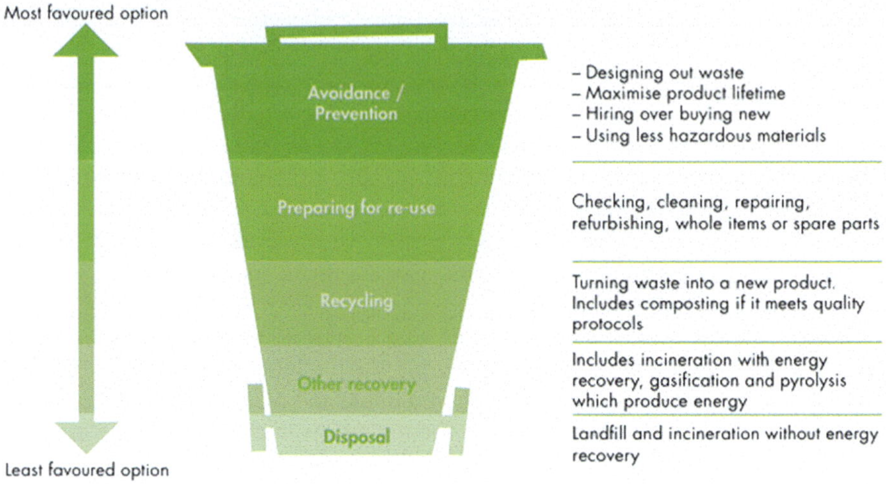


**Fig. 7** Waste management system [19]

hazardous materials be considered. Such items should be reusable after cleaning or repairing without changing its properties [19]. The next step of the waste management system recommends that if a product cannot be reused in its present form, then it should atleast be recyclable, such that its building material can be used to fabricate a new and possibly different product [1].

The recovery level of the system encourages that if a product cannot be fully recycled, then we should seek to recover the components and convert them into something usable. If there are no components of the product that can recover, and must indeed be disposed, then the first approach should be using it as fuel. Lastly if the waste product cannot fit in any of the above levels, then it must be disposed accordingly, in landfills or any other accredited and well monitor disposal plots [1].

## 4 Converting Waste into Energy Storage

As discussed above, the waste content is detrimental to our environment and human health. It is therefore imperative that we avoid generating waste or recycle the waste material that we have generated. The ladder can benefit us financially, if carried out effectively. Not only will it improve our financial position, but will assist in minimizing pollution, preserving our environment and will preserve our well-being and that of the marine life.

Different types of technologies can be developed from different types of waste materials. In my previously reported chapters, it was shown that sea shell waste could be recycled to generate luminescent materials, which could be used to fabricate self-glowing light bulbs, and other luminescent products [13]. It was also shown that different devices could be fabricated from organic waste material, such as electronic chips, solar cells and energy storages [21].

In this chapter, my interest is in showing that some materials can be recycled and converted into active electrodes of a Lithium ion battery.

## 4.1 Understanding a Lithium Ion Battery

A Lithium-ion battery is a rechargeable type of a battery, which derives energy from an electrochemical system. Such a battery is made of two electrodes: the Cathode and the Anode [22]. The Cathode is made up of a Li-compound, which can be $LiCoO_2$, $LiMn_2O_4$, $LiNiO_2$ or $LiFePO_4$, just to mention a few [22, 23]. The corresponding anode material is mostly made up of carbon material [23]. The Cathode and the Anode are separated by an electrolyte material (Fig. 8), which is a neutral medium that enables ion movement from the cathode to the anode [22, 23]. $LiPF_2$, $LiClO_4$ and $LiAlCl_4$ are among the commonly used electrolyte mediums in the lithium-ion battery production [24].

To charge a Lithium-ion battery, a positive potential difference is placed across the cathode and the anode. It forces the positive lithium ions to flow from the cathode

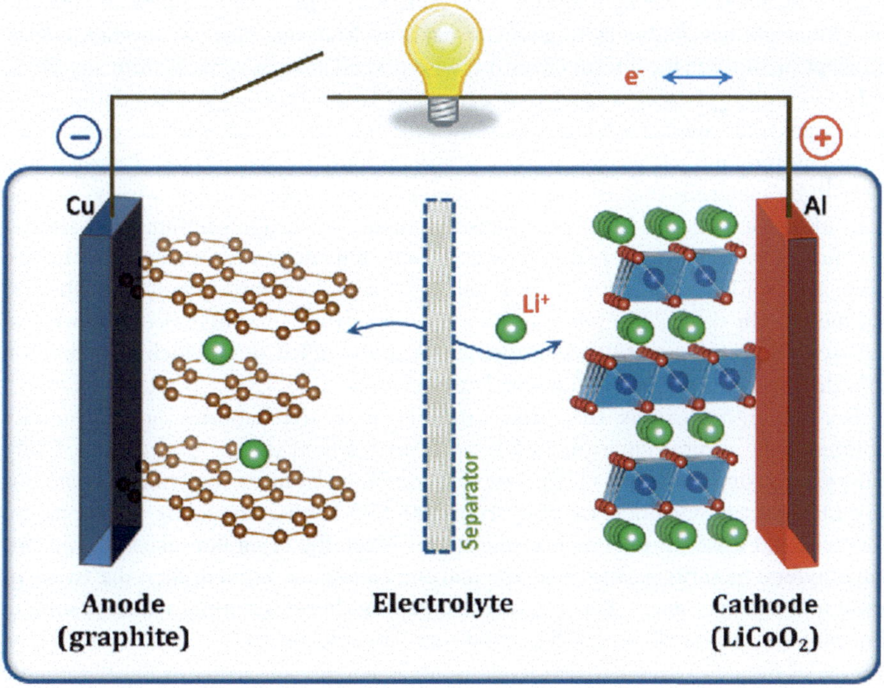


**Fig. 8** Schematic of a lithium ion battery [22]

to the anode via the electrolyte medium, and the corresponding electrons flow from the cathode to the anode via the external circuit. When the anode is fully saturated, then the flow of Lithium ions and electrons comes to a stop, and this is an indication that the battery is fully charged [25]. When a device that uses a battery is connected to it, it supplies a negative potential across the electrodes, and therefore activates the opposite movement of lithium ions and electrons. The electron movement in the external circuit ensures energy supply to the paired device [25].

### *4.2 Waste Material Converted to a Cathode Electrode Materials*

The Cathode material is the carrier of the energy of the battery, because is it electron rich [22]. As indicated above, several Lithium compounds are used to fabricate cathode material. The materials used to fabricate the cathode are often natural minerals, which are synthesized to create the product. However, this is not the only route, because the waste materials can also be recovered or recycled and used to prepare an active cathode of material. In this section we shall focus on preparation of $LiCoO_3$ cathode from a waste battery material, with an inactive cathode electrode.

Our daily lives are surrounded by technology that is aimed at making our lives easier, which is powered by batteries. This includes smartphones, laptops and electric cars. Smartphones enable us to navigate our way from one place to another, source information through the internet from anywhere, send business emails from anywhere and they enable us to access services in a reasonably short period, just to mention a few advantages. the batteries which support these technological gadgets pose great threat to our environment when disposed carelessly in dumping sites. This is because they are built by materials that are hazardous and possess a lot of energy, which can lead to accidental explosions [26]. The electrolyte material used in Lithium-Ion batteries is sufficient to corrode any material (Fig. 9), which it encounters under the influence of the air, should it leak from a cell. It can also badly contaminate the water, the soil and the human skin [27]. Cobalt in its ionic form ($Co^{2+}$) has also been reported to negatively affect human [29]. This element is found in $LiCoO_3$, which is a common cathode material used in Lithium-ion batteries [23].

Accidental explosions from the Lithium-ion based batteries are commonly reported, and several smartphones have been recalled after sales, due to faulty batteries, causing explosions [30]. The cathode of the Lithium-ion batteries contains very high energy densities, ranging from 200–690 Wh/L, depending on the nature of compound used for the cathode electrode. When the separator, usually a porous polyethylene membrane, of the anode and the cathode is broken, then the two will come into direct contact. This leads to dangerous short circuiting, which results in heat building up inside a cell. The heat increases the pressure, which leads to the rapture of the cell container and leads to air leakage. This leads to the flammable electrolyte igniting and causing a serious explosion due to the high energy density

**Fig. 9** Corroded lithium-ion cells affecting the environment [28]

of the cathode material [25, 27]. These explosions and the released gases can harm human beings and nearby animals.

Due to the hazardous nature of the heavy metals present in Lithium-ion batteries and the corresponding accidental explosions (Fig. 10), which have been reported, regulations have been put into place to safely handle and dispose of the batteries. Among the disposal methods is recycling of the battery building materials. Different cathode materials can now be recycled and re-used again in a new battery [26].

Several cathode materials can now be recovered through recycling and be re-used in developing new batteries. Several researchers have shown how to recycle an inactive $LiCoO_2$ cathode material using different chemical processing methods [31–33]. In a recent study, the lithium content was recycled to $Li_2CO_3$ through NaOH precipitation. The recovery of Cobalt has been carried out through precipitation using oxalic acid to form $CoC_2O_4$ [31], using NaOH to form $Co_3O_4$ [32] and using $H_2SO_4$ to form $CoSO_4$ [33]. The newly formed compounds are used as reagent compounds to form $LiCoO_2$ material that is readily active to be used as a cathode for the lithium-ion batteries [31–33].

Figure 11 shows a schematic process that was followed to recover lithium and cobalt from waste battery material using hydrothermal reaction [34]. The initial stage is dismantling the lithium-ion battery cells to expose the electrode material. The electrodes are separated from the aluminum foil used as a current collector [31, 32, 34]. The aluminum foil is treated with NaOH and recovered as $Al_2O_3$ [32]. After the powder of $LiCoCo_2$ has been scraped off the aluminum foil, it is heat treated and then filtered with an acid from the graphite content.

**Fig. 10** A burnt smart phone after a Lithium-ion battery explosion [31]

The cobalt and lithium ions are separated by precipitation using NaOH and $Na_2CO_3$ [32] or $H_2SO_4 + H_2O$ via a hydrothermal reaction, as indicated in Fig. 11 [34]. This results in the formation of $Co_3O_4$ and $Li_2CO_3$, which can be used as reagents in an oxygenated environment to synthesize $LiCoO_2$ via solid state reaction method [31, 32]. Figure 12 shows the XRD pattern of the $LiCoO_2$ phase, of the newly recycled compound from a used waste and inactive lithium ion battery, which is now active and ready to be used as a cathode material [32].

### *4.3 Waste Material Converted to an Anode Electrode Material*

The anode of the Lithium-ion batteries is essential for the storage of the Lithium ions, when the battery is charged. It therefore determines the storage capacity of the battery, which is an indication of the amount of current that can be transported to the external circuit in time. A material that has been conventionally used as an anode electrode for the Lithium-ion batteries is graphite, because of its capacity to trap Lithium ions and its chemical stability. However, several other materials are now being researched for the anode material to improve the capacity of batteries, which include graphene [35].

With graphene able to accept almost double the amount of Lithium ions than can be accepted by the conventional graphite anode, it therefore becomes a better candidate material for the future anode electrode materials. This greatly proves to increase the

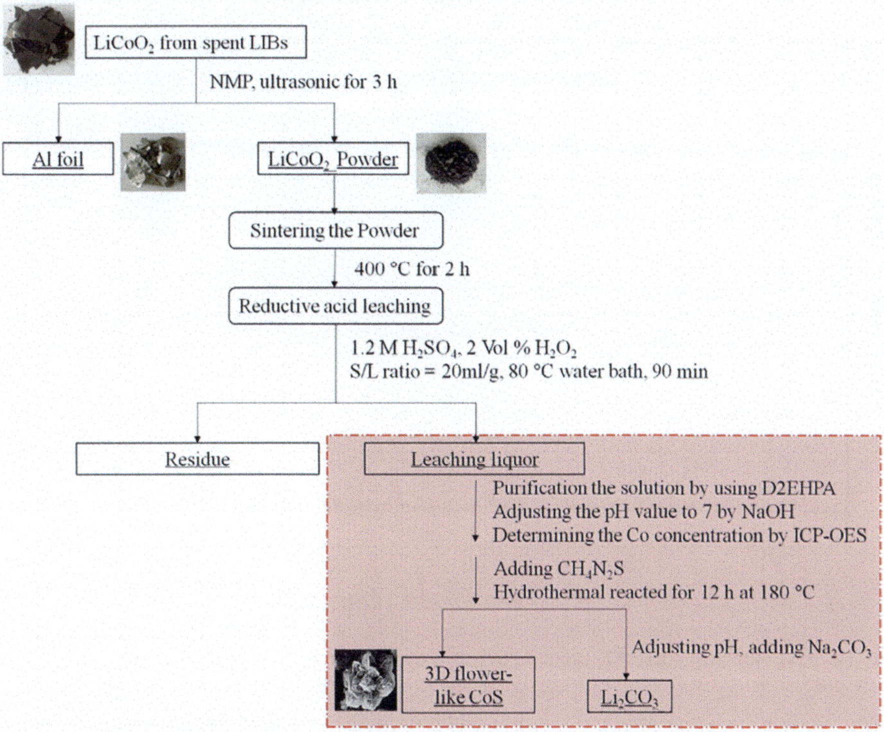


**Fig. 11** Method to recover lithium and cobalt from the waste Lithium-ion battery [34]

electrical conductivity anode, leading to an increased prolonged discharge of Lithium ion from the anode. This translates to a battery that will take long to discharge. The graphene material stores the Lithium ion migrating from the cathode by adsorption onto its atomic layer, and which can be prepared into a graphene foam that is rich with nanopores (Fig. 13) [36].

Graphene is conventionally prepared by two methods; the top-down and he bottom-up method. The top-down method involves pilling down a bulky carbon material into slices of graphene sheets, using mechanical or chemical exfoliation. The bottom-up method involves combining carbon atoms to assemble a molecule in two dimensions, forming a graphene sheet through different chemical synthetic routes [21].

Several methods to fabricate graphene from organic waste, which includes plants, plastic and insects have recently been discovered [21, 38]. This makes the fabrication of graphene to be cost effective. To explore the method used to prepare graphene from plants, below we explore the usage of brown rice husks to fabricate graphene compound [39].

The method involves converting brown rice husks into a carbon compound by burning it at 500 °C to a black ash (Fig. 14), then add potassium hydroxide to the

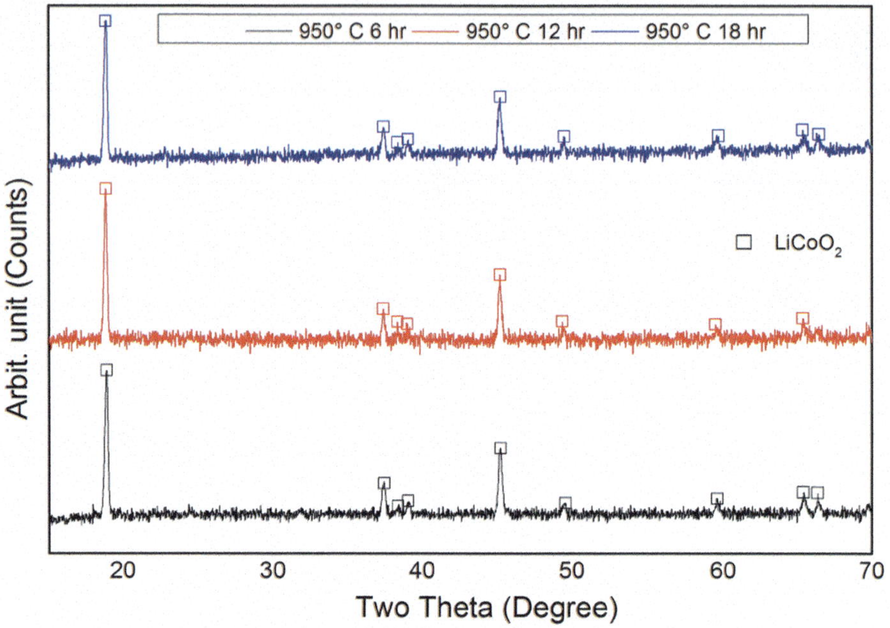


**Fig. 12** XRD spectra of $LiCoO_2$ reclaimed from a used waste lithium ion battery [32]

ash. The two are grounded to a fine mixture, which is later annealed at 700 °C and under the following processes [39]:

$$6KOH + 2C(Ash)\ \overrightarrow{700\,^{\circ}C}\ 2K + 3H_2 + 2K_2CO_3,$$

$$2K_2CO_3 + SiO_2(Ash)\ \overrightarrow{700\,^{\circ}C}\ C + K_2SiO_3 + O_2$$

$$K_2SiO_3 + C + H_2O\ \overrightarrow{RT}\ C + (K_2SiO_3 + H_2O)solution$$

100 ml of deionized water was added into the activated compound and stirred for several hours. The potassium compound formed a solution, and the carbon content saturated at the bottom as the residue and was extracted by filtering the solution. The process was repeated several times, to ensure successful removal of the potassium compound. Finally, the compound was dried at 150 °C for 24 h to obtain the graphene nanosheets, from the brown rice husks [39]. The Scanning Electron Microscope image (Fig. 15) of the final compound was captured, and it confirmed the formation of graphene nanosheets [39]. This new compound is now ready to be used in preparation of an anode electrode. This proves that the brown rice husks, which were initially going to be thrown away as waste, can be converted into a component of an important product.

**Fig. 13** An image of graphene foam with nanopores captured by an electron microscope [37]

## 5 Batteries Recharging the Economy

The new technologies which are electricity driven strongly rely on batteries to store their energy (Fig. 16), leading to batteries being their backbone [40]. The Lithium-ion batteries are designed in different sizes and shapes, ranging from coin cells, cylindrical cells, pouch cells, to stack or pack cells into batteries of different sizes.

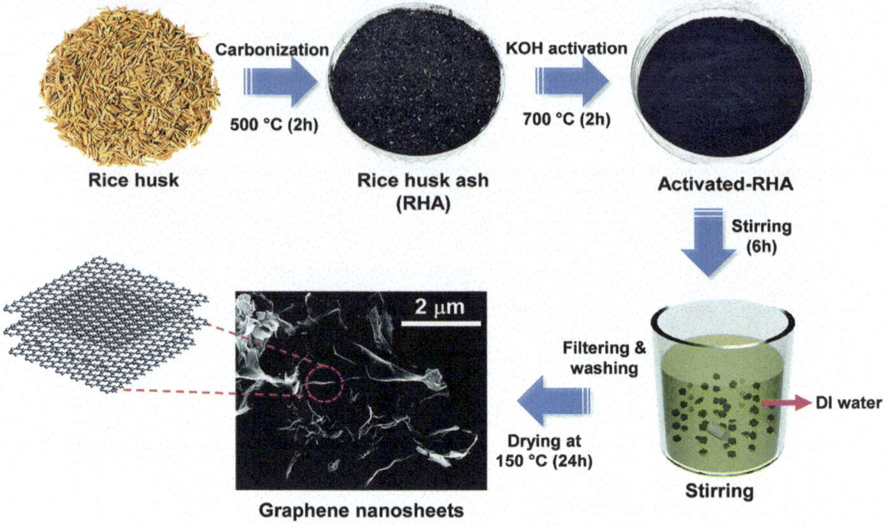


**Fig. 14** A schematic showing rice husk used to fabricate graphene [39]

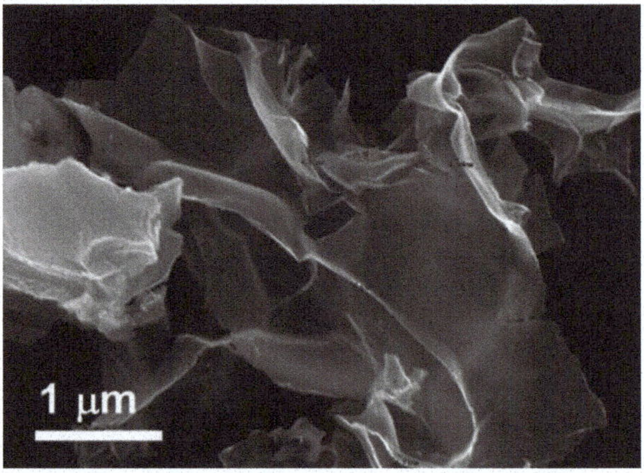


**Fig. 15** Scanning electron microscope image of a graphene sheet [39]

This makes them suitable in supporting different types of technologies such mobile phones, computers, power tools, electric cars, home micro-grids storages and Photovoltaic solar plants, just to mention a few (Fig. 16) [41]. The number of cars powered by electricity is also increasing, due to several countries pledging their support to reduce global warming [40].

Apple is among the biggest technological companies, globally [42], and its technologies rely on batteries to store energy. An article published by business day on the 2nd of August in 2018, suggests that the market capitalization of Apple is bigger than that of the Gross Domestic Product of South Africa, Netherlands, Switzerland and many more other countries [43]. According to Forbes magazine, the company

**Fig. 16** A list of possible applications for the lithium ion battery [41]

employs more than half a million employees [44], The number excludes the private shopping outlets that sell apple devices. The backbone of Apple devices is the battery technology. Not only are batteries economically supporting Apple, but several other companies like Tesla, Panasonic and Samsung, just to mention a few [45]. Batteries even support cellphone communication, which result in several business deals and marketing. Without a doubt, batteries greatly support the economic growth of many countries, if not all.

## 6 Conclusion

In our busy daily lives, we find ourselves throwing away everything that is not of need to us, without thinking twice. However, we know that most of those items can be recycled and converted into products or devices that can support the economy of several countries. Most of the waste products that we find ourselves throwing away, end up polluting our own environments and lead to sickness and difficult living conditions. In this work I have shown that several waste materials can be recycled and converted into active electrodes of a Lithium-ion battery. These batteries greatly support our daily lives and the economic growth of companies, and countries thrive on the usage of devices or products that are supported by batteries. This urges everyone, before throwing away something that looks like waste, to think of different means of reusing, recovering or recycling that item, and turn it into something that can generate profit. If not for profits and economic benefits, then ensure a safe environment from pollution.

## References


1. South African Department of Environmental Affairs. (2012). *National waste information baseline report*. Pretoria.
2. Galvez, R. P., & Berge, J. P. (2013). *Utilization of fish waste*. CRC-Press.
3. Retrieved November 19, 2018, from http://soleco-technology.com/what-we-do/sewage-waste-water/.
4. Dharmappa, H. B., Sivakumar, M., Singh, R. N. (1995). *Wastewater minimization and reuse in mining industry in Illawarra region*. International Mine Water Association.
5. Johnson, B. (2013). *Zero waste home: The ultimate guide to simplifying your life by reducing your waste*. Scribner.
6. Retrieved November 19, 2018, from https://image.pennlive.com/home/penn-media/width620/img/news/photo/2017/02/20/recycled-pilejpg-4ffde1298190a58a.jpg.
7. Retrieved November 19, 2018, from https://mpumalanganews.co.za/313921/ngomane-says-community-part-liquor-licenses-approval/.
8. Polprasert, C. (2007). *Organic waste recycling: Technology and management* (3rd ed.). IWA Publishing.
9. Retrieved November 20, 2018, from https://www.planetnatural.com/composting-101/making/what-to-use/.
10. Retrieved November 20, 2018, from http://www.moreno-valley.ca.us/resident_services/waste/hazardous-waste.shtml.
11. Retrieved November 20, 2018, from https://www.health24.com/Medical/Cholera/Recent-cholera-outbreaks/Recent-cholera-outbreaks-in-SA-20120721.
12. Conner, N. (2009). *Living green: The missing manual*. O'Reilly Media Inc.
13. Noto, L. L., Mofokeng, S. J., Sintwa, N., Mochane, M. J., Mothudi, B. (2018). Deriving luminescent materials from sea shell waste. In L. Z. Linganiso (Ed.), *Waste-to-Profit (W-t-P): Value added products to generate wealth for a sustainable economy* (Vol. 1). New York: Nova Science publishers Inc.
14. Hofer, T. N. (2008). *Marine pollution: New research*. Nova Science publishers Inc.
15. Retrieved November 20, 2018, from https://www.metropolitantransferstation.com.au/blog/negative-effects-of-improper-waste-management.

16. Duarte, A. C., Cachada, A., & Rocha-Santos, T. A. P. (2018). *Soil pollution: From monitoring to remediation*. Academic Press.
17. Retrieved November 21, 2018, from https://timesofindia.indiatimes.com/city/madurai/Excess-plastic-waste-in-landfills-kills-soil-fertility-Eco-experts/articleshow/48027259.cms.
18. Weart, S. R. (2003). *The discovery of global warming*. Harvard University Press.
19. Retrieved November 22, 2018, from https://www.recycledevon.org/at-work/resource-efficiency.
20. Bortoleto, A. P. (2015). *Waste prevention policy and behaviour: New approaches to reducing waste generation and its environmental impacts*. Routledge.
21. Molefe, F. V., Noto, L. L., Mbongo, M., Mochane, M. J., Dhlamini, M. S. (2018). Organic waste converted into energy producing and storing systems. In L. Z. Linganiso (Ed.), *Waste-to-Profit (W-t-P): Value added products to generate wealth for a sustainable economy* (Vol. 1). New York: Nova Science publishers Inc.
22. Goodenough, J. B., & Kyu-Sung, P. (2013). The Li-ion rechargeable battery: A perspective. *Journal of the American Chemical Society, 135*(4), 1167.
23. Mishra, A., Mehta, A., Basu, S., Malode, S. J., Shetti, N. P., Shukla, S. S., et al. (2018). Electrode materials for lithium-ion batteries. *Materials Science for Energy Technologies, 1*, 182.
24. Jow, T. R., Xu, K., Borodin, O., & Ue, M. (2004). *Electrolytes for lithium and lithium-ion batteries*. Springer Media.
25. Warner, J. T. (2015). *The Handbook of lithium-ion battery pack design: Chemistry, components, types and terminology*. Elsevier Inc.
26. Wang, L. K., Hung, Y. T., & Shammas, N. K. (2017). *Handbook of advanced industrial and hazardous waste treatment*. CRC Press.
27. Levy, S. C., & Bro, P. (1994). Battery hazards and accident prevention. New York: Plenum Press.
28. Retrieved December 05, 2018, from https://mdmurbach.github.io/cei-demo-tutorial/lithium-ion-afterlife.html.
29. Leyssens, L., Vinck, B., Van Der Straten, C., Wuyts, F., & Maes, L. (2017). Cobalt toxicity in humans–a review of the potential sources and systemic health effects. *Toxicology, 387*, 43.
30. Retrieved December 05, 2018, from https://www.consumerreports.org/safety-recalls/why-lithium-ion-batteries-still-explode-and-whats-being-done-to-fix-the-problem/.
31. Jian, G., Guo, J., Wang, X., Sun, C., Zhou, Z., Yu, L., et al. (2012). Study on separation of cobalt and lithium salts from waste mobile-phone batteries. *Procedia Environmental Sciences, 16*, 495.
32. Qabir, R., & Gulshan, F. (2018). Reclamation of Lithium Cobalt oxide from waste lithium ion batteries to be used as recycled active cathode materials. *Materials Sciences and Applications, 9*, 142.
33. Mao, J., Li, J., & Xu, Z. (2018). Coupling reactions and collapsing model in the roasting process of recycling metals from $LiCoO_2$ batteries. *Journal of Cleaner Production, 205*, 923.
34. Chen, H., Zhu, X., Chang, Y., Cai, J., & Zhao, R. (2018). 3D flower-like CoS hierarchitectures recycled from spent $LiCoO_2$ batteries and its application in electrochemical capacitor. *Materials Letters, 218*, 40.
35. Mekonnen, Y., Sundararajan, A., Sarwat, A. I. (2016). A review of cathode and anode materials for lithium-ion batteries. In *Proceedings of the 2016 SoutheastCon* (Vol. 2). VA, USA: Norfolk.
36. Raccichini, R., Varzi, A., Passerini, S., & Scrosati, B. (2015). *Nature Materials, 14*, 271.
37. Xiao, X., Michael, J. R., Beechem, T., McDonald, A., Rodriguez, M., Brumbach, M. T., et al. (2012). Three dimensional nickel–graphene core–shell electrodes. *Journal of Materials Chemistry, 22*, 23749.
38. Ruan, G., Sun, Z., Peng, Z., & Tour, J. M. (2011). Growth of graphene from food, insect and waste. *ACS Nano, 5*(9), 7601.
39. Sankar, S., Lee, H., Jung, H., Kim, A., Ahmed, A. T. A., Inambar, A. I., et al. (2017). *New Journal of Chemistry, 41*, 13792.
40. Retrieved January 24, 2019, from https://www.theguardian.com/environment/2015/dec/12/paris-climate-deal-200-nations-sign-finish-fossil-fuel-era.

41. Retrieved December 03, 2018, from http://www.stbchip.co.kr/en/contents/product/lithium-ion-battery.htm?sm=2_2_1.
42. Galloway, S. (2017). *The four: The hidden DNA of Amazon, Apple, Facebook and Google*. Random House Large Print.
43. Retrieved December 20, 2018, from https://www.businesslive.co.za/bd/companies/telecoms-and-technology/2018-08-02-10-amazing-facts-about-apple-which-just-became-a-trillion-dollar-company/.
44. Retrieved January 24, 2019, from https://www.forbes.com/sites/connieguglielmo/2012/03/02/apple-touts-itself-as-big-job-creator-in-the-u-s/#468ac0d81a60.
45. Retrieved January 24, 2019, from https://www.visiongain.com/report/top-20-lithium-ion-battery-manufacturing-companies-2018/.

# Ethanol Production from Various Wastes

**Tshwafo E. Motaung, Setumo V. Motloung, and Lehlohonolo F. Koao**

**Abstract** Bioethanol is a promising alternative for gasoline as it is produced mainly from corn and sugarcane. However, alternative sources of corn starch and cellulose are required since both corn and sugarcane are primary sources of food. Landfill capacities are shrinking due to accumulated waste that requires utilisation into value added products for effective waste management systems and reduction of environmental pollution problems due to shortage of waste disposal areas. Moreover, fossil fuels emit a high number of greenhouse gases which are another global concern. Producing bioethanol from waste reduces GHG emissions and environmental pollution. This chapter focuses on utilisation of waste such as municipal solid waste, lignocellulosic feedstock, food waste into bioethanol as conversion of waste to energy and valuable chemicals. This is achieved by pre-treatment techniques, hydrolysis and fermentation.

**Keywords** Bioethanol · Sugarcane bagasse · Waste management · Pre-treatements

## 1 Introduction

In a world where fossil fuels are depleting at a high rate, alternative and non-petroleum based energy sources are a necessity [1]. Some of the main energye sources are highlighted in Fig. 1. The transportation sector utilises 60% of the world petroleum-based fuels, which makes it accountable for the highest emission of greenhouse gases (GHG), responsible for the emission of 70% carbon monoxide (CO) and 19% carbon

T. E. Motaung (✉)
Department of Chemistry, Sefako Makgatho Health Sciences University, P.O. Box 94, Medunsa 0204, South Africa
e-mail: tshwafo.motaung@smu.ac.za; motaungte@live.com

S. V. Motloung
Central University of Technology, Private Bag X20539, Bloemfontein 9300, South Africa

L. F. Koao
Department of Physics, University of the Free State (Qwa Qwa Campus), Private Bag X13, Phuthaditjhaba 9866, South Africa

 
M. Tshwafo Elias (ed.), *Waste to Value Innovations*,
https://doi.org/10.1007/978-981-96-2782-0_6

dioxide ($CO_2$) [2]. Moreover, South Africa has the highest GHG emission in Africa [3]. This affects the global ecosystem stability and is responsible for global climate change. This sector is in need of alternative bio-based fuels (biofuels), to reduce the strain on the depletion of fossil fuels and the reduction of GHG emissions globally [1] Biofuels are bio-based materials from the conversion of chemical energy stored by biomass during photosynthesis into energy rich gases and liquids [4]. They have a potential to be alternatives of fossil fuels based materials due to similar properties [5]. Biofuels are bio-based materials including bioethanol, biohydrogen, bio-oil, vegetable oils, biogas, biodiesel, bio-synthetic gas, biochar, and Fischer–Tropsch fuel [2]. Only 0.8% biofuels are currently used compared to fossil fuels (78.3%) [6].

Biofuels are classified as first generation biofuels, second generation biofuels, third generation biofuels and fourth generation biofuels. The first generation biofuels are unsustainable and obtained from corn, wheat, cassava, barley, rye, soybean, sugarcane, sugar beet, or sweet sorghum which are rich in sugar and starch [7–9]. The second generation of biofuels are sustainable and can be obtained from non-food crops, solid waste, and lignocellulosic feedstock. The third generation biofuels are obtained from aquatic animals; algae and sea weeds. The fourth generation biofuels is obtained from vegetable oils [10] (Table 1).

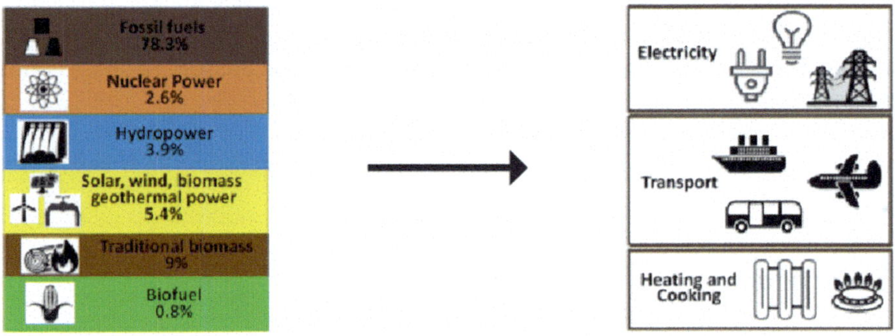


**Fig. 1** Energy sources [6]

**Table 1** Classification of biofuels by generation based on their feedstocks [5]

| Generation | Feedstocks | Examples |
|---|---|---|
| First-generation biofuels | Sugar, starch, vegetable oils, or animal fats | Bioalcohols, vegetable oil, biodiesel, biosyngas, biogas |
| Second generation biofuels | Non food crops, wheat straw, corn, wood, solid waste, energy crop | Bialcohols, bio-oil, bio-DMF, biohydrogen, bio-Fischer–Tropsch diesel |
| Third generation biofuels | Algae | Vegetable oil, biodiesel |
| Fourth generation biofuels | Vegetable oil, biodiesel | Biogasoline |

## 2 Properties of Ethanol

Bioethanol (ethyl alcohol EtOH or chemically $C_2H_5OH$) is a liquid oxygenated biofuel that can be used as an alternative gasoline replacement due to its high-octane number (108) [2], high heat of vaporization, high flame speed, and broad flammability limit in comparison to gasoline [2, 5]. It could reduce GHG emissions by reducing $CO_2$ build-up as it gets recycled during combustion as the source of production is non-fossil fuel based [2]. However, bioethanol is miscible in water compared to gasoline. The other disadvantages are lower energy density, flame luminosity is low, and vapor pressure is lower than gasoline. It is also corrosive.

Bioethanol is produced by microorganisms during saccharification and fermentation of sugars (cellulose and hemicellulose) or starch. The fermentable sugars have chemical energy which is converted to renewable energy sources [1, 4]. Primary biofuels (such as wood) are used unprocessed, while secondary biofuels (bioethanol and biodiesel) are processed prior to usage in transportation and petrochemical industries [11]. Bioethanol can be used as a transportation fuel for cars operating on ethanol or anhydrous ethanol, can be blended with gasoline or diesel to improve physical and chemical properties, and for the reduction of exhaust gases emission [2, 5]. In fact, this addresses the transportation request of ethanol, as it has been globally in demand due to its potential to be used as a fuel, and a synthesis of chemicals and polymers [10]. Furthermore, it is a chemical feedstock for ethylene industrial applications. Ethylene is used to produce polyethylene and other polymers [10, 12]. Ethanol is also a precursor in the generation of green $H_2$ for fuel cells applications [13]. Globally, large-scale biofuel production is more costly than the production of petroleum fuels. About two-thirds of the production costs of bioethanol are from the feedstock, which increases the competition between bioethanol production and food production [5, 11]. However, this does not obliterate the economic advantages of the production of biofuels such as sustainability, reduction of GHG emissions, value added bio-based products and regional development, improving the strength of rural development and agricultural economy [14, 15]. In addition to the mentioned economic advantages of producing bioethanol, there will be an increase in employment rate since several opportunities will arise, such as farming opportunities, plant operations, equipment manufacturing, and market opportunities for domestic farmers [14, 15]. There are two common bioethanol synthesis routes, namely, biological treatment of feedstock and the hydration of ethylene [10]. Both bioethanol production routes are employed at industrial scale. Biological pre-treatment is the conversion of sugars extracted from feedstock through fermentation to recover ethanol and is known to be a conventional process. The second route involves ethylene obtained from fossil fuels to produce ethanol in catalysis [10]. There are various steps in bioethanol production dependent on the source, namely, pre-treatment including hydrolysis, fermentation, and the recovery of ethanol which is both distillation and dehydration.

## 3 Pre-treatment of Feedstock

Complexity of different feedstock requires pre-treatment methods prior to fermentation due to the various components in the structural composition such as cellulosic materials [1]. There are several types of pre-treatment methods that are used for lignocellulosic biomass deconstruction. These methods are categorized into physical, chemical and physicochemical, and biological pre-treatment methods [16, 17]. Physical methods include separations, milling, irradiation, and liquid hot water, pyrolysis and pulse-electric-field also known as ultrasound [17]. Biological method is a pre-treatment technique that employs usage of a metabolic microorganism to degrade both lignin and hemicellulose during the pre-treatment. The metabolism of these microorganisms, which are white, brown or soft-rot fungi which enhances the enzymatic saccharification rate during the pre-treatment of biomass [18]. Chemical and physicochemical methods are alkaline pre-treatment, bleaching, organosolv, ionic liquids, microwave-chemical pre-treatment, acid hydrolysis, ammonia fibre explosion (AFEX), $CO_2$ explosion and supercritical fluid pre-treatment [19–21]. All pre-treatment techniques have advantages and disadvantages over others, in most cases, combination of two or more techniques is more effective compared to using one technique [22]. The composition and different ultrastructure of plant cell walls of bioenergy plants require different pre-treatment techniques based on the resulting sugars and sugar conversion in bioethanol production [7]. Hydrolysis and fermentation are the next step conducted after pre-treatment process for producing bioethanol. During hydrolysis, different inhibitors such as furfural and hydroxymethylfurfural (HMF) are formed. These by-products hinder the yield of ethanol produced from starch and cellulose due to sugar losses [23]. Detoxification is one way of reducing the inhibitors formed during acid hydrolysis of starch and cellulose.

## 4 Fermentation

The conversion process of pentose and hexose sugars into bioethanol using microorganisms, such as bacteria, yeast, or fungi is known as fermentation [24]. Bioethanol conversion processes are separate hydrolysis and fermentation (SHF), simultaneous saccharification and fermentation (SSF), simultaneous saccharification and co-fermentation (SSCF), simultaneous saccharification and juice co-fermentation (SSJcF), and consolidated bioprocessing step (CBP) [11, 25] SHF produces poor ethanol yield due to the separated pretreatment process and saccharification processes [11, 25]. During this process, monosaccharide sugar is produced, followed by fermentation. However, the simultaneous saccharification and fermentation techniques, whereby fermentation and enzymatic hydrolysis occur simultaneously, produce high ethanol yield due to low residual sugar after the process. During this process, hydrolysis and fermentation happen simultaneously, where glucose is directly converted to ethanol [26]. Moreover, it favors the production of ethanol as it improves the reaction rates, yields and ethanol concentration [27].

## 5 Carbohydrates Used in Ethanol Production

Sugar and starch rich food crops have been used to produce bioethanol due to the chemical energy stored by biomass during photosynthesis. The chemical energy is then bio-converted during fermentation process into energy rich biofuels which are renewable energy sources [4, 5]. Cellulose and starch are carbohydrates required to produce bioethanol. They are hydrolysed to sugars before fermenting to ethanol. Thus, the interest in corn (*Zea mays* L.) and sugarcane (*Saccharum officinarum* L.) which is the feedstock used in the production of ethanol, has now been criticized due to food vs fuel issues in developing countries [28]. Corn fermentation to ethanol was more technically advanced compared to cellulosic fermentation to ethanol. However, over the years, the ethanol production gap has been reduced and both sources are efficient for ethanol production [29]. The production of ethanol from both corn and sugarcane is shown in Fig. 2.

Starch rich feedstock also requires pre-treatment before fermentation of sugars [28]. The starch is first pre-treated by a gelatinization process, then liquefaction process which involves heat and enzymes, and either pressurized or unpressurised conditions for an accelerated process. The last step being the fermentation process, where the sugars are converted to ethanol. Alternatively, simultaneous saccharification and fermentation techniques can be used for an effective process [28].

## 6 Sources of Bioethanol Production

The utilisation of cheap resources, such as lignocellulosic biomass, solid wastes, and starches/sugars as a source of bioethanol production will reduce the conflicts of food versus fuel while reducing the GHG emissions and improving sustainability [2, 11, 22]. Lignocellulosic biomass includes agricultural crops and residues, forest residues, energy feedstock, food processing wastes, milling wastes, garden residues [22]. Solid wastes include paper waste, municipality solid waste (MSW), non-recycled paper and recycled material. Sources of starches and sugars are agricultural crops, i.e., sugarcane, wheat, maize, rice, etc., and food processing residues rich in starches and sugars [22]. Tabah et al. [31] converted starch to ethanol in a simultaneous saccharification and fermentation process which yielded ethanol that was tested in fuel cells as a potential fuel source. One reaction step for the successful conversion of starch to ethanol showed a possible technique that requires no additional energy input in ethanol production.

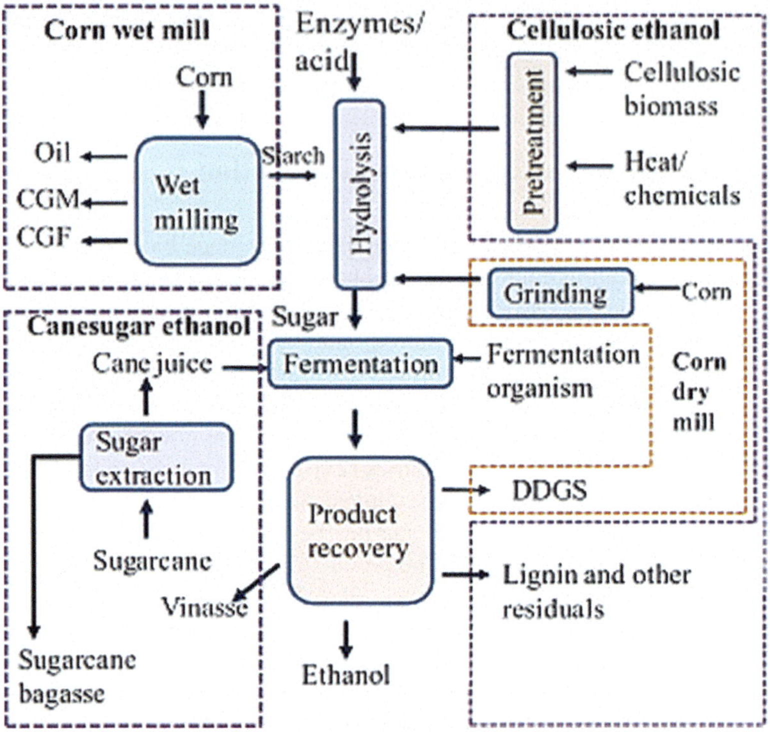


**Fig. 2** Bioethanol production from corn and sugarcane (canesugar) [30]

## *6.1 Foodstuff Waste*

Globally, one-third of food produced for human consumption is lost or wasted along the food supply chain and end up in landfills or incinerated [32]. Innovative uses of food waste which includes the production of bioethanol are necessary as means to manage food waste by utilizing the readily available source of biofuels [32]. Food waste includes food lost during production, storage, transportation, and processing stages of the food supply chain, and waste food found in restaurants and common households [33]. About 13% of SA general waste is organic food waste, of which 65% is disposed to landfills [33]. Food waste has been successfully used as a raw material in the production of bioethanol as it is rich in renewable organic and nutrient contents [34]. Food waste contains cellulose and starch, it requires less pre-treatment or no pre-treatment before saccharification compared to lignocellulosic residues [35] (Fig. 3).

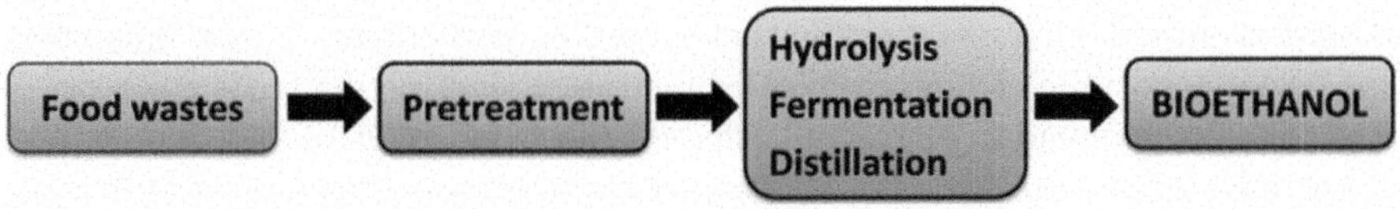


**Fig. 3** Conversion of food waste to ethanol process [35]

Dhiman et al. [36] recently produced bioethanol during SHF of unprocessed food waste using thermophilic anaerobic bacteria coupled with high-temperature-high-pressure distillation techniques with hope to eliminate pre-treatment and enzymatic hydrolysis. Bioethanol was successfully produced, however there were noted complications that deserve further research before piloting the method.

## *6.2 Municipal Solid Waste (MSW)*

Municipal solid waste has potential to be utilised as feedstock for bioethanol production due to the availability in landfills, thus, causing land pollution which results in occupation of fertile space and health problems [23]. Producing biofuel from municipal solid waste improves waste management sustainability and reduces the amount of waste disposed of in landfills. Wastepaper is part of the municipal solid waste, and it is the largest constituent in the waste. It can be used to produce bioethanol due to the presence of cellulose. Wastepaper contains mostly cellulose and a portion of lignin depending on the source [37]. Wastepaper is a rich cellulosic feedstock that has been used to produce bioethanol since it is cheaper, and the production of ethanol from wastepaper reduces GHG emissions compared to corn [38]. This will assist in vacating landfills and render space available for an innovative activity to manage waste further. Cellulose present in wastepaper was converted to sugars through hydrolysis, which was then followed by fermentation of sugars to ethanol [38]. The amount of carbohydrates present was 70%, with a lignin of 16%. Pre-treatment of the waste was done to remove lignin which and separate hemicellulose from cellulose. This improves effectiveness of hydrolysis and fermentation processes during the conversion of reducing sugars to bioethanol [38].

## *6.3 Lignocellulosic Feedstock*

Lignocellulosic materials are mainly comprised of three abundant natural polymers which are cellulose, hemicellulose which are both carbohydrates, and the second most abundant natural polymer, lignin. There are also extraneous materials like ash and extractives present [23]. For lignocelluloses materials, it usually needs pretreatment and hydrolysis to convert cellulose and hemicellulose into structural sugar monomers [26]. The composition of lignocellulosic biomass depends on the feedstock type.

Yusuf and Inambao [24] recently studied the production of bioethanol from different *Matooke* peels species. *Matooke* is an indigenous green cooking banana species found from Southwest Uganda, which is among the highest producers of bananas in the world [24]. The fruit can be edible upon cooking, while the peels are either used as feedstock or discarded in landfills. It has a high starch content compared to ordinary bananas, which makes the *Matooke* peels suitable to produce bioethanol. *Matooke* peels are ±35.0% weight per banana fruit and consists of ±70% carbohydrates [24]. The production of bioethanol was a success with 90% ethanol produced. Zhang et al. [25] recently produced ethanol from sweet sorghum juice mixed with sweet sorghum bagasse using simultaneous saccharification and juice co-fermentation (SSJcF). The sweet sorghum bagasse was pre-treated with diluted acid, and the sweet sorghum juice was extracted from crushed sweet sorghum stalks. Wheat straw and corn stover were used to produce ethanol through SSCF technique. The lignocellulosic feedstock was pre-treated before SSCF removed lignin and free hemicellulose. Xylose and cellulose were then converted to sugars and resulted in high titer of ethanol [29].

## 7 Conclusion

Corn, sugarcane and sugar beets are the main sources of bioethanol. However, as primary value agricultural crops for food, there seems to be a challenge to fully utilize it as the main source of bioethanol because this could be a threat to food security. Alternative feedstock to produce bioethanol such as lignocellulosic biomass, solid waste, and starches/sugars as a source of bioethanol production will reduce the food versus fuel debate while reducing the GHG emissions and improving sustainability of the environment. Bioethanol is an alternative gasoline transportation fuel that can also be used in fuel cells. It forms a very small portion of energy sources which need to change. Starch rich and cellulosic feedstock can be used to produce bioethanol after pre-treatment, hydrolysis, fermentation and distillation. Most ethanol production from waste feedstock is only limited to laboratory scale. More pilot scale technology is required for implementation of waste to bioethanol as an energy source. Properly combined pre-treatment processes must increase the fermentable sugars concentrations before fermentation. Others, efficient microbes and genetically modified microbes may also enhance enzymatic hydrolysis.

## References

1. Kim, J. H., Lee, J. C., & Pak, D. (2011). Feasibility of producing ethanol from food waste. *Waste Management, 31*, 2121–2125.

2. Balat, M. (2011). Production of bioethanol from lignocellulosic materials via the biochemical pathway: A review. *Energy Conversion and Management, 52*, 858–875.
3. Jain, S., & Jain, P. K. (2017). The rise of renewable energy implementation in South Africa. In *World Engineers Summit–Applied Energy Symposium & Forum: Low Carbon Cities & Urban Energy Joint Conference, WES-CUE 2017, 19–21 July 2017* (Vol. 143, pp. 721–726). Singapore: Energy Procedia.
4. Güell, E. J., Maru, B. T., Chimentão, R. J., Gispert-Guirado, F., Constantí, M., & Medina, F. (2015). Combined heterogeneous catalysis and dark fermentation systems for the conversion of cellulose into biohydrogen. *Biochemical Engineering Journal, 101*, 209–219.
5. Demirbas, A. (2009). Political, economic and environmental impacts of biofuels: A review. *Applied Energy, 86*, S108–S111.
6. Voloshin, R. A., Rodionova, M. V., Zharmukhamedov, S. K., Nejat Veziroglu, T., & Allakhverdiev, S. I. (2016). Review: Biofuel production from plant and algal biomass. *International Journal of Hydrogen Energy, 41*(39), 17257–17273.
7. Balan, V. (2014). Current challenges in commercially producing biofuels from lignocellulosic biomass. *ISRN Biotechnology, 2014*, 463074–463074.
8. Acheampong, M., Ertem, F. C., Kappler, B., & Neubauer, P. (2016). In pursuit of sustainable development goal (SDG) number 7: Will biofuels be reliable?
9. Carneiro, T. F., Timko, M., Prado, J. M., & Berni, M. (2016). Chapter 17-biomass pretreatment with carbon dioxide. In S. I. Mussatto (Ed.), *Biomass fractionation technologies for a lignocellulosic feedstock based biorefinery* (pp. 385–407). Amsterdam: Elsevier.
10. Sikarwar, V. S., Zhao, M., Fennell, P. S., Shah, N., & Anthony, E. J. (2017). Progress in biofuel production from gasification. *Progress in Energy and Combustion Science, 61*, 189–248.
11. Sharma, N., & Sharma, N. (2018). Second generation bioethanol production from lignocellulosic waste and its future perspectives: A review. *International Journal of Current Microbiology and Applied Sciences, 7*(5), 1285–1290.
12. Kiran, E. U., & Liu, Y. (2015). Bioethanol production from mixed food waste by an effective enzymatic pretreatment. *Fuel, 159*, 463–469.
13. Subramani, V., & Song, C. (2007). Advances in catalysis and processes for hydrogen production from ethanol reforming. *Catalysis, 20*(6).
14. Demirbas, A. (2007). Producing and using bioethanol as an automotive fuel. *Energy Sources, Part B, 2*(4), 391–401.
15. Demirbas, A. (2017). The social, economic, and environmental importance of biofuels in the future. *Energy Sources, Part B: Economics, Planning, And Policy, 12*(1), 47–55.
16. Lee, H., Hamid, S., & Zain, S. (2014). Conversion of lignocellulosic biomass to nanocellulose: structure and chemical process. *The Scientific World Journal, 2014.*
17. Akhtar, N., Gupta, K., Goyal, D., & Goyal, A. (2016). Recent advances in pretreatment technologies for efficient hydrolysis of lignocellulosic biomass. *Environmental Progress & Sustainable Energy, 35*(2), 489–511.
18. Sindhu, R., Binod, P., & Pandey, A. (2016). Biological pretreatment of lignocellulosic biomass–an overview. *Bioresource Technology, 199*, 76–82.
19. Zheng, Y., Lin, H.-M., Wen, J., Cao, N., Yu, X., & Tsao, G. T. (1995). Supercritical carbon dioxide explosion as a pretreatment for cellulose hydrolysis. *Biotechnology Letters, 17*(8), 845–850.
20. Brandt, A., Ray, M. J., To, T. Q., Leak, D. J., Murphy, R. J., & Welton, T. (2011). Ionic liquid pretreatment of lignocellulosic biomass with ionic liquid–water mixtures. *Green Chemistry, 13*(9), 2489–2499.
21. Brodeur, G., Yau, E., Badal, K., Collier, J., Ramachandran, K. B., & Ramakrishnan, S. (2011). Chemical and physicochemical pretreatment of lignocellulosic biomass: A review. *Enzyme Research, 2011.*
22. Kumari, D., & Singh, R. (2018). Pretreatment of lignocellulosic wastes for biofuel production: A critical review. *Renewable and Sustainable Energy Reviews, 90*, 877–891.
23. Karimi., S., & Karimi, K. (2018). Efficient ethanol production from kitchen and garden wastes and biogas from the residues. *Journal of Cleaner Production, 187*, 37–45.

24. Yusuf, A. A., & Inambao, F. L. (2018). *Bioethanol production from different Matooke peels species: A surprising source for alternative fuel.* Case Studies in Thermal Engineering.
25. Zhang, C., Wen, H., Chen, C., Cai, D., Fu, C., Li, P., Qin, P., & Tan, T. (2019). Simultaneous saccharification and juice co-fermentation for high-titer ethanol production using sweet sorghum stalk. *Renewable Energy, 134*, 44–53.
26. Dahnum, D., Tasum, S. O., Triwahyuni, E., Nurdin, M., & Abimanyu, H. (2015). Comparison of SHF and SSF processes using enzyme and dry yeast for optimization of bioethanol production from empty fruit bunch. *Energy Procedia, 68*, 107–116.
27. Wyman, C. E., Spindler, D. D., & Grohmann, K. (1992). Simultaneous saccharification and fermentation of several lignocellulosic feedstocks to fuel ethanol. *Biomass and Bioenergy, 3*(5), 301–307.
28. Szambelan, K., Nowak, J., Szwengiel, A., Jeleń, H., & Łukaszewski, G. (2018). Separate hydrolysis and fermentation and simultaneous saccharification and fermentation methods in bioethanol production and formation of volatile by-products from selected corn cultivars. *Industrial Crops and Products, 118*, 355–361.
29. Liu, G., Zhang, Q., Li, H., Qureshi, A. S., Zhang, J., Bao, X., & Bao, J. (2018). Dry biorefining maximizes the potentials of simultaneous saccharification and co-fermentation for cellulosic ethanol production. *Biotechnology and Bioengineering, 115*(1), 60–69.
30. Vohra, M., Manwar, J., Manmode, R., Padgilwar, S., & Patil, S. (2014). Bioethanol production: Feedstock and current technologies. *Journal of Environmental Chemical Engineering, 2*(1), 573–584.
31. Tabah, B., Pulidindi, I. N., Chitturi, V. R., Arava, L. M. R., & Gedanken, A. (2015). Solar-energy driven simultaneous saccharification and fermentation of starch to bioethanol for fuel-cell applications. *ChemSusChem, 8*(20), 3497–3503.
32. Wunderlich, S. M., & Martinez, N. M. (2018). Conserving natural resources through food loss reduction: Production and consumption stages of the food supply chain. *International Soil and Water Conservation Research, 6*(4), 331–339.
33. de Lange, W., & Nahman, A. (2015). Costs of food waste in South Africa: Incorporating inedible food waste. *Waste Management, 40*, 167–172.
34. Zhang, Z., O'Hara, I. M., Mundree, S., Gao, B., Ball, A. S., Zhu, N., Bai, Z., & Jin, B. (2016). Biofuels from food processing wastes. *Current Opinion in Biotechnology, 38*, 97–105.
35. Karmee, S. K. (2016). Liquid biofuels from food waste: Current trends, prospect and limitation. *Renewable and Sustainable Energy Reviews, 53*, 945–953.
36. Dhiman, S. S., David, A., Shrestha, N., Johnson, G. R., Benjamin, K. M., Gadhamshetty, V., & Sani, R. K. (2017). Simultaneous hydrolysis and fermentation of unprocessed food waste into ethanol using thermophilic anaerobic bacteria. *Bioresource Technology, 244*, 733–740.
37. Nishimura, H., Tan, L., Kira, N., Tomiyama, S., Yamada, K., Sun, Z.-Y., Tang, Y.-Q., Morimura, S., & Kida, K. (2017). Production of ethanol from a mixture of waste paper and kitchen waste via a process of successive liquefaction, presaccharification, and simultaneous saccharification and fermentation. *Waste Management, 67*, 86–94.
38. Dubey, A. K., Gupta, P. K., Garg, N., & Naithani, S. (2012). Bioethanol production from waste paper acid pretreated hydrolyzate with xylose fermenting Pichia stipitis. *Carbohydrate Polymers, 88*(3), 825–829.

# Energy Funding in South Africa

**Tshwafo E. Motaung**

**Abstract** Waste has contributed to global warming issues globally. It has become one of the major environmental challenges faced by the world today. Waste to energy is advantageous in the utilisation of waste into renewable energy sources. However, the commercialization of renewable energy initiatives is costly compared to the traditional fuel-based energy. There is still a barrier in financing the renewable energy projects which can be alternatives to fossil fuels. This chapter explores the South African funding for energy related projects and initiatives. There is The Green Fund, Commercial Banks, Development Finance Institutions, South African Government Departments, and other renewable energy institutions, and Private Equity companies which offer different funding for energy initiatives. Each funding institution has its own requirements and offer either a grant or a loan, or both.

## 1 Introduction

The annual consumption of goods that have a short lifespan and that only requires single usage increases the amount of waste generated globally. The increase in the amount of waste generation and waste composition is dependent on the consumption patterns, and each country's consumption patterns, industrial and economic structures [1]. The global population is estimated to be 8 billion in the year 2025, and 9.3 billion in the year 2050 from the current world population of an estimated 7 billion. The annual increase in the fertility rate is estimated at 80 million, which is still rising [2] (Fig. 1).

Both hazardous and non-hazardous waste is produced by human activities [3]. About 1.3 billion tons of waste, mainly municipal solid waste (MSW) is globally generated per annum. It has also been estimated that this value will rise up to 2.2 billion tons by the year 2025 [1]. In South Africa, waste management involves collection of waste, disposal of waste in landfills and recycling of waste, which is

T. E. Motaung (✉)
Department of Chemistry, Sefako Makgatho Health Sciences University, P.O. Box 94, Medunsa 0204, South Africa
e-mail: tshwafo.motaung@smu.ac.za; motaungte@live.com

 
M. Tshwafo Elias (ed.), *Waste to Value Innovations*,
https://doi.org/10.1007/978-981-96-2782-0_7

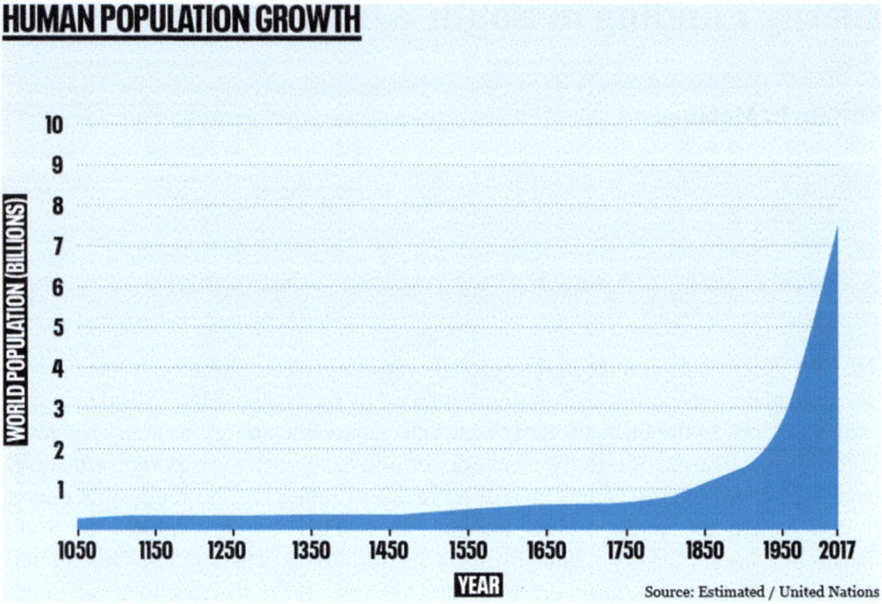


**Fig. 1** Global human population growth [2]

currently limited [4–6]. Waste definition is different among countries, but waste can be described as any substances or objects that will eventually be disposed or discarded [1]. The disposed waste can be reused, recycled, or discarded in landfills. However, in Japan, any substances or objects that can be recycled is not considered waste since they still have value. The European Union Waste Framework Directive (2008) Waste hierarchy is reduction of waste, reusing waste, recycling waste, recovery of waste and dumping waste in landfills, and incineration [3, 7, 8] (Fig. 2).

As the landfills run out of space, waste recycling is the most critical step in waste management as it has a potential to save energy, increases the preservation of natural resources, is effective in the reduction of greenhouse gases, and hinders the leakage of toxins from landfills to the environment [8]. Incineration is not environmentally friendly option for waste management. In South Africa, waste generated is managed by local municipalities and private companies. Local municipalities collect and manages waste generated in households, while private companies collect and manages commercial and industrial waste [9]. About 90% of waste is disposed in municipality landfills [9–11]. South African landfills are 90% occupied with waste, there is still a gap in waste management services [11]. South Africa has identified waste management as crucial for the green economy and for the sustainability of the environment by reforming policy and regulations regarding waste, acknowledging that the landfill space is limited, and increasing awareness to the citizens [12, 13].

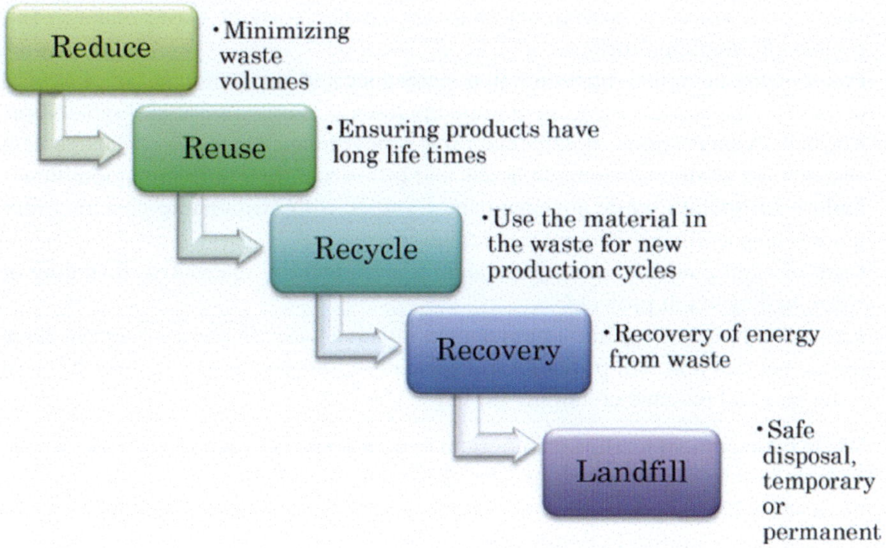


**Fig. 2** The hierarchy of waste adopted from Nehrenheim [3]

Waste management can improve South African economic growth and create sustainable jobs in which The National Green Economy Accord [14] has recognised as a transition towards a South Africa's Green Economy [15–17].

The Financial and Fiscal Commission, under the Money Bills Amendment Procedure and Related Matters Act (No. 9 of 2009), undertakes research and makes annual recommendations regarding national raised revenues and equitable sharing of revenues [18]. The Financial and Fiscal Commission is responsible for the waste management financial mechanisms in South Africa. In the numerous waste management studies that has taken place in South Africa, which have been focused on operational management, environmental aspects and the legislative framework, there has been fewer studies in the waste management funding mechanisms and policies in South Africa [18]. The Financial and Fiscal Commission has a number of critical policy issues and key factors in relation to financing waste management in South Africa as listed as the following points:

- "The implications of classifying waste management as a free basic service (FBS).
- The lack of a standardised approach for setting appropriate and equitable tariff structures for waste management services.
- 'Ring fencing' funds from the Municipal Infrastructure Grant (MIG) for capital expenditure on specific infrastructure.
- The distribution of funding for waste management across the three tiers of government—(national, provincial and local government).
- Ambiguous functions, powers and institutional arrangements across government, which affect access and budgetary control of waste management projects.

- Poor optimisation of operational and maintenance costs of waste management projects in municipalities.
- Poor costing of capital, operational, maintenance and environmental implications of waste management projects in municipalities.
- The lack of and/or poor implementation of proper financial and asset management systems for waste management assets and infrastructure within municipalities.
- Failure to operate waste management services with sustainable cost recovery models, especially in indigent municipalities.
- Lack of clear guidance on sources and requirements for alternative financing of waste management projects.
- Lack of will to develop and implement innovative waste management projects in municipalities that will provide for job creation, improved service delivery, cost reduction and revenue creation." [9, 18].

Waste management funding is important in South Africa's green economy growth. It is also essential for Research and Development (R&D) in academic institutions and industries [15–17]. This chapter focuses on the availability of funding of waste and energy opportunities in South Africa.

## 2 Energy Opportunities in SA

African countries have rather very minimal accurate and reliable waste management information. South Africa also has the same problems regarding waste issues such as quantifying the amount of waste (tonnages) produced [19]. The waste sector had an estimation of R 15.3 billion gross domestic product (GDP) in 2012. A report by the National Waste Information Baseline (NWIB) showed that in 2011, South Africa waste generation was approximately 108 million tonnages [20] Thus, 59 million tonnages was general waste, 48 million tonnages was unclassified waste, and 1 million tonnages was hazardous waste [19]. About 65% of the above estimated classified waste is recyclable, which means it should be recovered from the dumping sites to be reprocessed [19, 20]. Furthermore, South Africa's waste management sector is involved in waste recycling, reuse, waste treatment and waste disposal [19].

## 3 Funding Institutions

There is funding available for supporting green businesses in South Africa. The information is available in the Green Cape which has classified the funding opportunities as The Green Finance Desk database [21]. Funding sources and incentives are described in this section.

a. **The Green Fund**

The Green Fund is a South African national fund that supports South Africa's green transitioning to low carbon, efficiently resourced and a development climate resilient country which delivers growth in green socioeconomic and environmental benefits.

b. **Commercial Banks**

Commercial banks offer loans for energy related businesses, both small and well established businesses. The amounts covered by the loans differ from each bank and the payment terms and conditions also differ.

i. **Absa**

"Absa Bank Limited (Absa Bank), with preference shares listed on the JSE Limited, is a wholly owned subsidiary of the Barclays Africa Group. We offer a range of retail, business, corporate and investment banking, and wealth management products and services primarily in South Africa and Namibia." [21]. Absa partnered with the French Development Agency (AFD) to provide funding for energy related projects for up to R 100 million as a loan. Their investment duration is a minimum of 3 years [21].

ii. **First National Bank (FNB)**

"As a financial institution, FNB provides personal, private, business, commercial and corporate banking services to millions of customers across South Africa." [21]. FNB provided private project financing and credit for renewable energy businesses. Their loan differs from business to business.

iii. **GreenFin Financial Services**

"GreenFin is a privately funded company, founded in 2012. GreenFin provides Green Energy Finance Solutions at preferential interest rates." [21]. GreenFin offers solar installation loans for homes and for Small and Medium Enterprises (SMEs).

iv. **GroFin**

"GroFin is a pioneering SME development financier assisting entrepreneurs and business owners to access tailored finance and experienced business support to start and grow successful businesses, creating jobs and driving socio-economic development." [21]. Financially supporting/loaning all sectors of small and growing businesses in Africa and the Middle East.

v. **Investec**

"Investec's provides a diverse range of financial products and services to a niche client base in three principal markets, the United Kingdom, South Africa and Australia, as well as certain other geographies." [21]. Investec Power & Infrastructure Finance offers loans for energy related projects.

vi. **Merchant West**

"Merchant West is an Authorised Financial Services Provider with a difference. We specialise in business finance and advisory services and strive to bring about flexible, innovative solutions, customised to our clients' needs. Our Products: For over a decade, Merchant West has been steadily growing its business across the financial services sector. An independent investment holding company, we offer specialised business finance and advisory services." [21]. Merchant West offers loans to solar energy related projects with an amount of up to R 25 million. They cover capital and short-term requirements.

vii. **Nedbank**

"Nedbank Group is one of South Africa's four largest banks, with Nedbank Limited as our principal banking subsidiary."[21]. Nedbank offer loans that depends on the renewable energy project. They are also partnered with AFD.

viii. **Rand Merchant Bank**

"A leading African Corporate and Investment Bank and part of one of the largest financial services groups (by market capitalisation) in Africa—FirstRand Limited." [21]. Rand Marchant Bank offer loans for businesses.

ix. **Sasfin**

"Sasfin is a banking and financial services group, offering tailor-made products and services to entrepreneurs, corporates and high net-worth individuals." [21]. Sasfin offers grants and loans to entrepreneurs, corporates and high net-worth individuals to fund their businesses.

x. **Standard Bank**

"Standard Bank is a leading Africa-focused financial services organisation. Our strategy is to build the leading Africa-focused financial services organisation using our competitive advantage to the full." [21]. Standard Bank partnered with AFD to offer loans for energy related businesses.

c. **Development Finance Institutions (DFI) or Development Banks or Development Finance Company (DFC)**

i. **Development Bank of South Africa and South South North**

"At South South North we advocate business unusual transitions to low carbon, climate resilient economies that seize the opportunities of sustainable development for the global South. South South North finds evidence-based, locally tailored solutions to climate and development challenges, in partnership with other leading organisations. South South North contributes to global knowledge in order to achieve climate compatible development in practice."[21]. The South South North and the Development Bank of Southern Africa (DBSA) manages the fund which aids with scaling up energy related development business in South Africa. The fund is operational under the Sustainable Settlements Facility (SSF).

ii. **Energy and Environment Partnership**

"The Energy and Environment Partnership (EEP) is a Programme which promotes renewable energy (RE), energy efficiency (EE), and clean technology investments. The EEP Programme supports projects which aim to provide sustainable energy services to the poor and combated the climate change. EEP provides seed financing to cover part of the project costs which are necessary to start and develop a business (such as pilot and demonstration activities) or which can create something in value so that it is worthy of investment (such as pre-feasibility and bankable feasibility studies)." [21] The grant is up to a maximum of R 5 million. The funding is offered when the projects with high innovation in energy services, facilitating technology transfer, encouraging cooperation and local stakeholders are close to being commercialised as bridging funds [21]. Biomass, waste-to-energy, biogas, solar PV, wind power, hydro and hybrid, are their main focus areas.

iii. **Renewable Energy and Energy Efficiency Partnership (REEEP)**

"REEEP invests in clean energy markets in developing countries to reduce $CO_2$ emissions and build prosperity. Based on a strategic portfolio of high impact projects, REEEP works to generate energy access, improve lives and economic opportunities, build sustainable markets, and combat climate change." [21]. REEEP offers a grant to clean energy projects both in public and private sector.

iv. **The African Development Bank**

"Grant resources and technical assistance to private project developers/promoters to finalize pre-investment activities." [21]. Private sector energy funding for African projects for renewable energy and energy efficiency.

d. **South African Government Departments**

i. **Department of Energy (DoE)**

"The Department of Energy is responsible for ensuring exploration, development, processing, utilisation and management of South Africa's mineral and energy resources. As the country's economy continues to grow, energy is increasingly becoming a key focus." DoE funds Energy Efficiency Demand Side Management (EEDSM) Programme through the Division of Revenue Act (National Treasury). DoE offers grants to energy projects that are managed by municipalities.

ii. **Eskom**

"Eskom generates approximately 95% of the electricity used in South Africa and approximately 45% of the electricity used in Africa. Eskom generates, transmits and distributes electricity to industrial, mining, commercial, agricultural and residential customers and redistributors. Eskom will continue to focus on improving and strengthening its core business of electricity generation, transmission, trading and distribution." [21]. Eskom offers Integrated Demand Management (IDM) programme, which grants a rebate on the purchase and installation of selected energy efficiency/renewable energy products.

iii. **Industrial Development Corporation (IDC)**

"Established in 1940, we are a national development finance institution set up to promote economic growth and industrial development. We are owned by the South African government under the supervision of the Economic Development Department." [21]. IDC partnered with the German Development Bank (KfW) to offer The Green Energy Efficiency Fund for energy efficiency and renewable energy projects in South Africa, which are privately owned. IDC also partnered with AFD to finance Green businesses and municipal projects. IDC does not offer grants but a loan.

iv. **South Africa National Energy Development Institution (SANEDI)**

"SANEDI's main function is to direct, monitor and conduct applied energy research and development, demonstration and deployment, as well to undertake specific measures to promote the uptake of green energy and energy efficiency in South Africa." [21]. SANEDI offers a 12L EE Tax Incentive: Tax incentives are being introduced for businesses that can show measurable energy savings.

v. **South African Revenue Services (SARS)**

"The South African Revenue Service (SARS) is the nation's tax collecting authority. Established in terms of the South African Revenue Service Act 34 of 1997 as an autonomous agency, we are responsible for administering the South African tax system and customs service." [21]. SARS offers a 12B RE Depreciation Incentive: Under section 12B of the Income Tax Act, businesses can depreciate investments in renewable energy and bio fuel production at a rate of 50:30:20.

**Other energy funding institutions**

vi. **Renewable Energy Performance Platform (REPP)**

"REPP works to mobilise private sector development activity—and investment—in small to medium-sized projects (typically up to 25 MW), and is supported with £148 m funding from the UK's Department for Business, Energy and Industrial Strategy (BEIS) and the International Climate Finance initiative." [21]. REPP offers a development phase capital, gap financing, grants risk to mitigation instruments and grants long-term landing access. This type of grant is a loan.

vii. **SolarSol**

"At SolarSol we specialise in providing Solar PV that makes financial sense. Our overall objective is not only to provide a quality solar PV installation but to also ensure that the cash savings provided by the installation generates an acceptable investment return." [21]. SolarSol provided a loan for the business that install solar systems.

e. **Private Equity**

i. **Actis**

"Actis is a leading investor in growth markets across Africa, Asia and Latin America. Actis is a global platform offering a multi-asset strategy through the asset classes of private equity, energy, infrastructure and real estate. We invest with an exclusive focus on intra-growth market investing (our 'south-south' approach): sharing knowledge and experience across regions and businesses to help deliver consistent, competitive returns responsibly." [21]. Actis funds energy related businesses.

ii. **African Infrastructure Investment Managers (AIIM)**

"African Infrastructure Investment Managers (AIIM) develops and manages private equity infrastructure funds designed to invest long-term institutional unlisted equity in African infrastructure projects. AIIM actively manages investments in East, West and Southern Africa and has equity under management of USD 2.1 billion with a track record extending across seven African infrastructure funds." [21]. They offer grants for renewable energy (solar and wind) projects under the IDEAS Fund.

iii. **Finvest Blue**

"Finvest Blue assists businesses in reaching new financial heights, through expert financial advice, financial modelling, reporting and compliance services. The entrepreneurial spirit of Finvest Blue is what ensures that we offer our clients more than just the normal accounting and tax services. We facilitate business partnerships at all stages of growth, we also assist in developing business models and consult on growth strategies." [21]. Finvest provides funding for South African Solar PV projects that are larger than 200 kWp and are willing to sign a 20-year Power Purchase Agreement.

iv. **Inspired Evolution**

"Inspired Evolution is a specialised investment management business and authorised financial services provider located principally in South Africa. Inspired Evolution offers a dedicated team with a deep global track record positioned to lead clean energy and resource efficiency investments across sub-Saharan Africa." [21]. Inspired Evolution has a funding under the name of Evolution One Fund that provides private equity for innovative clean energy projects, both public and private sectors in the Southern African Development Community.

v. **Persistent Energy Capital (PEC)**

"Persistent Energy Capital (PEC) focuses on providing off-grid households with access to basic energy services: electric light, mobile phone charging, fans, radios, TV and Internet. PEC has invested in and incubated businesses with a commercial approach to providing such energy services to low-income customers in sub-Saharan Africa. PEC also provides financial and strategic advice to investors, governments and others participating in the development of distributed renewable energy sector." [21]. PEC provides energy funding as loans for private sector and community-based organisations.

vi. **TBI**

"Established in 1992, TBI is a South African alternative investment and fund management group holding company, with a history of superior returns. The business consists of a number of direct investments in fund management and related associates (Awande Investment Managers, Bridge Capital and TBI Investment Managers), and three core investment portfolios." [21]. TBI offer renewable energy independent power producer procurement (REIPPP) projects, financing or refinancing where they have existing debt to equity holders. Credit Suisse is the financial provider of the funding for REIPPP projects.

vii. **Triumph Venture Capital**

"Triumph Venture Capital (Pty) Ltd. is a private equity and venture capital firm specializing in start-up, early stage, expansion stage, and growth capital financing. It seeks to invest in the technology companies based in South Africa." [21]. Triumph Venture Capital funds renewable energy sector by providing growth capital, corporate acquisitions and recapitalisations.

## 4 Conclusion

South Africa waste to energy initiatives needs more funding for the projects to contribute to green economy and increase job opportunities. South Africa has a high unemployment rate, in which more businesses can operate through receiving funding from different funding institutions will be successful. There is The Green Fund, Commercial Banks, Development Finance Institutions, South African Government Departments, and other renewable energy institutions, and Private Equity companies which offer different funding for energy initiatives. There are grants offered and loans by different funding types. Commercial banks offer loans, while Development Finance Institutions and Government Departments offer grants. There are different types of funding depending on the level of the business, and also whether the business operates in public or in private. South Africa has opportunities with the availability of different funding institutions that will enable businesses to strive and create jobs.

## References

1. Kawai, K., & Tasaki, T. (2016). Revisiting estimates of municipal solid waste generation per capita and their reliability. *Journal of Material Cycles and Waste Management*. (Springer)
2. Population Matters. (2019). Population: The numbers, population matters. https://populationmatters.org/the-facts/the-numbers?gclid=CjwKCAjwx_boBRA9EiwA4kIELqbMdwk8CeDj_t_4uaC7-MKwCAx8hlR0S5-ghV5qX6P3yDukx2ds6hoC1SUQAvD_BwE.
3. Nehrenheim, E. (2014). Waste management: Introduction. In *Reference module in earth systems and environmental sciences*. https://doi.org/10.1016/B978-0-12-409548-9.09166-1.

4. Marland, E., & Marland, G. (2003). The treatment of long-lived, carbon-containing products in inventories of carbon dioxide emissions to the atmosphere. *Environmental Science & Policy, 6*(2), 139–152. https://doi.org/10.1016/S1462-9011(03)00003-0.
5. Zhang, N., et al. (2011). Greening academia: Developing sustainable waste management at higher education institutions. *Waste Management, 31*(7), 1606–1616. https://doi.org/10.1016/j.wasman.2011.03.006.
6. Brunner, P. H., & Rechberger, H. (2015). Waste to energy–key element for sustainable waste management. *Waste Management, 37*, 3–12. (Elsevier)
7. Gharfalkar, M., et al. (2015). Analysis of waste hierarchy in the European waste directive 2008/98/EC. *Waste Management, 39*, 305–313. (Elsevier)
8. Gunther, M. (2007). *The end of garbage.* Fortune.
9. CSIR Paper. (2014). The state of innovation in the South African waste sector. In *Solid waste world congress.* Sao Paulo: ISWA 2014.
10. Nahman, A. (2011). Pricing landfill externalities: Emissions and disamenity costs in Cape Town, South Africa. *Waste Management.* https://doi.org/10.1016/j.wasman.2011.05.015.
11. Dada, O. R., & Mbohwa, C. (2016). Municipal solid waste from landfills a solution to energy crisis in South Africa. In *Lecture notes in engineering and computer science.*
12. Afful-Koomson, T. (2015). The green climate fund in Africa: What should be different? *Climate and Development.* https://doi.org/10.1080/17565529.2014.951015.
13. Schalatek, L., Bird, N., & Watson, C. (2017). *The green climate fund, climate finance fundamentals.*
14. Economic Development Department (EDD). (2011). *Economic Development Department Annual Report 2011–2012.* https://www.gov.za/sites/default/files/gcis_document/201409/eddannualreport20012012weba1.pdf
15. Borel-Saladin, J. M., & Turok, I. N. (2013). The impact of the green economy on jobs in South Africa. *South African Journal of Science.* https://doi.org/10.1590/sajs.2013/a0033.
16. African Centre for a Green Economy. (2015). *Mapping the green economy landscape in South Africa.* SA Green Fund.
17. Moyo, T. (2015). Local government and green jobs creation: Exploring opportunities in selected metropolitan municipalities in South Africa. *Journal of Public Administration.*
18. Madubula, N., Makinta, V. (2013). *Financing of waste management in South Africa.*
19. Business Sweden. (2017). *High potential opportunity waste to energy in South Africa.*
20. Department of Environmental Affairs. (2012). National waste information baseline report. Pretoria. http://sawic.environment.gov.za/documents/1880.pdf.
21. Green Cape. (2019). *Green finance databases, green cape.*

# Assessment of Energy Sector and Opportunities

**Tshwafo E. Motaung, Lehlohonolo F. Koao, and Setumo V. Motloung**

**Abstract** The world is largely powered by coal, oil, nuclear energy, renewables, natural gas, hydropower, petroleum, solar PVs, and wind power, with coal still the most dominant energy source globally. Current climate policies, economic sustainability, energy security, future global energy goals, sustainability goals, drive the transition from fossil fuels to renewables which are growing in shares of global energy productions, and projected to be dominant by 2050. For the first time in history, renewables have surpassed coal in year-on-year production growth, with solar PV capacity installations in 2017 alone levelling the total capacity the world had installed by 2012. The contribution of solar power to the global energy sector is projected to be above 20% by 2040. The lithium-ion battery storage, onshore wind, offshore wind, and utility-scale solar declined in cost by 2019 compared to 2018. Unfortunately, coronavirus (COVID-19) interrupted many sectors globally since the end of 2019. The widespread lockdowns lowered air pollution and $CO_2$ concentrations temporarily, though it increased the energy demand for the uses to curb the spread of the virus. The effect of the COVID-19 pandemic in the global energy sector involves the projected decline in investments in 2020, compared to 2019. Electricity, for the first time ever, is projected to become the biggest single element in energy consumption. The revenues of the power sector, including commodities are expected to fall. Most opportunities available in the global energy sector are investment based, on renewables, and deep electrification of the transport sector.

T. E. Motaung (✉)
Department of Chemistry, Sefako Makgatho Health Sciences University, P.O. Box 94, Medunsa 0204, South Africa
e-mail: tshwafo.motaung@smu.ac.za; motaungte@live.com

L. F. Koao
Department of Physics, University of the Free State (Qwa Qwa Campus), Private Bag X13, Phuthaditjhaba 9866, South Africa

S. V. Motloung
Central University of Technology, Private Bag X20539, Bloemfontein 9300, South Africa

 
M. Tshwafo Elias (ed.), *Waste to Value Innovations*,
https://doi.org/10.1007/978-981-96-2782-0_8

# 1 Introduction

## *1.1 The Global Energy Sector*

The global energy sector is largely dependent on coal, oil, nuclear energy, and renewables for energy supply. The transitioning from pro-fossil fuels to pro-renewables brought challenges and opportunities in the sector. Climate policies in place have driven some gradual shifts from fossil fuels, with many scenarios suggesting the global dominance of renewables in the next 3–4 decades. Many projections were made before COVID-19 which suggest increased error in the current data [1].

As of August 2020, there were about 19,303,517 confirmed cases of COVID-19 globally, causing crises in different sectors, including the global energy sector [2–4]. Various countries adopted and implemented different strategies to limit the spread of the virus, which partly impacted the global energy sector. During lockdowns, electricity demand was decreased by 20%, the electricity demand declined by 2.5% globally, in the first quarter of 2020, compared to 2019 [4]. The global largest electricity demand reduction was experienced in China, with 6.5%, whose lockdown began mid-January. Despite the decline in electricity demands, the global energy share used to contain the spread of COVID-19 hiked from 5% (mid-March) to 50% (mid-April), adding to other global energy demands and challenges, including $CO_2$ emission [2]. In 2018, it was argued that the shortage of electricity could reduce the probability of employability by 35 to 41%, which could be worse with the additional burden added by the pandemic [10.] In fact, the pandemic effects have taken a toll to investments. Investments in energy sector are negatively impacted by widespread lockdowns across countries and dip in revenues, especially for oil. The decline in investments in 2020 is predicted to be the biggest ever, by approximately USD 400 billion, compared to 2019. The global biggest energy consumer spending was Oil (50%) and electricity (a further 38%). The revenues of the power sector are expected to fall by USD 1 trillion, oil spending will drop by at least 180 billion in 2020, with a rise in non-payment. Another implication of these COVID-19 impacts on the sector is a historic change as electricity is projected to become the biggest single component of energy consumption spending [5]. The investments in crude oil and natural gas may be affected due to their respective decline shown in Fig. 2, across African countries [4].

## *1.2 Major Forms of Energy*

The global energy production is dominated by coal, oil, natural gas, nuclear, and renewables. During the period, 2000–2018, the global primary energy consumption increased by 40%, and by 2.1% from 2016 to the end of 2018. The consumed energy comprises oil (34.2%), coal (27.6%), natural gas (23.4%), hydropower (6.8%),

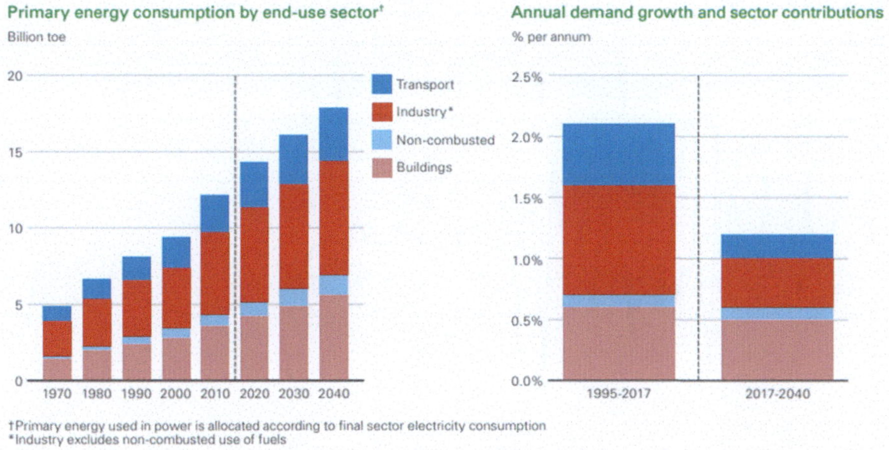


**Fig. 1** Projections of global primary energy consumption, annual demand growth, and sector contributions from 1970–2040, and 1995–2040, respectively [7]

nuclear (4.5%), and renewables (3.5%). Despite crucial investments made on renewable energy, fossil fuels are still dominant [6]. The growing liquefied natural gas (LNG) contributes to the growth of natural gas in the energy sector. During the period, 2019–2020, the global primary energy consumption & demand increased and projected to increase further until 2040, with sectors like transport, industry, and buildings (Fig. 1) [7]. Figure 1 displays model projections for the annual global primary energy consumption and demand growth and sector contributions until 2040. All sectors displayed in Fig. 1 show growth in contributions, in various scenarios.

The growing energy demand comes with challenges and opportunities in the sector which can be addressed through policies, technology, and investments in the promising energy sources [1] (Fig. 2).

## 2 Solar Power

Globally the solar PV capacities were the most installed power generation technologies in 2017. Solar power increased capacity to 50% more than renewable energy and wind power, fossil fuels and nuclear energy combined. In 2008, only 0.01% of global electricity was supplied by solar power, with above 2% growth by 2018. Projections suggest that the contribution of solar power to the global energy sector will be above 20% by 2040 [9]. The solar power growth in the global energy sector was 30% compared to 2016. The solar PV installations made in 2017 are equivalent to the total installations the world had made in total by 2012. China contributed 53.3% of the total global solar PV capacity installations made in 2017 and is projected to

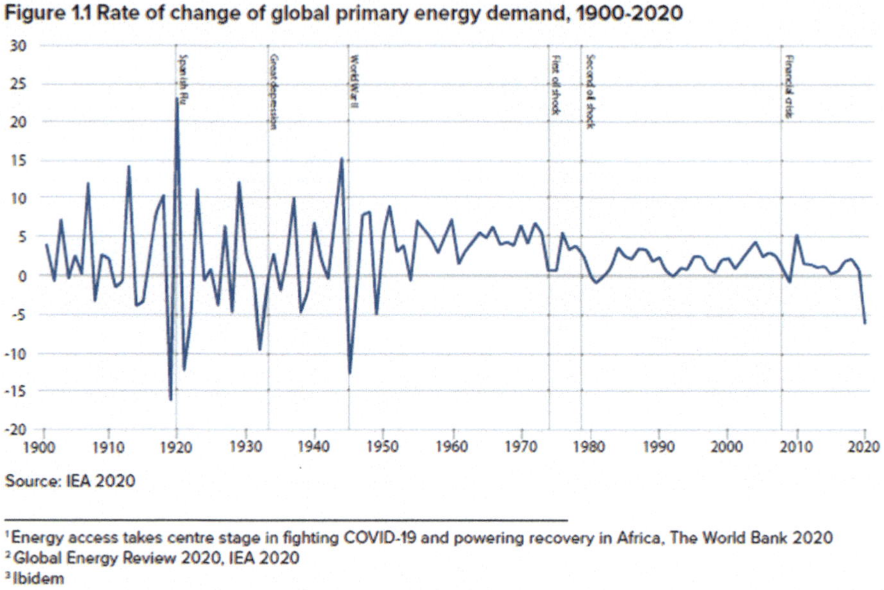


**Fig. 2** Rate of change of global primary energy demand, 1900–2020 [8]

be leaders by more than 50%, followed by North America (20%), and Europe (10%) by 2050, provided the capacities are scaled up significantly in the next 30 years.

Europe reported a 30% increase in solar PV installations, compared to 2016 [10]. According to some projections, solar PV can contribute 25% to the global total electricity by 2050. Investments in solar PV are critical to the projected increase to the total global energy sector by 2050, which is an opportunity. This would mean a 68% rise in annual solar PV investments until 2050, which in 2018 stood at USD 114 billion [1]. Technological improvements in solar PV capacities will continue to decrease the costs, making solar PV more economically viable [1, 11].

## 3 The Shifts and Global Outlook

Global trends indicate that renewables will be dominant in the future, with the projected decline in fossil fuels. Nonetheless, the global trend of energy production is changing, with the growth in production in the US, while the Chinese energy supplies are decreasing in expansion. The increase in renewables, oil, and gas drives the growth in production for the US, while the slowing production in China is attributed to their shift to sustainable economic growth. Besides slowing production, China has the largest growth in energy supplies attributed to the rapid growth of renewables, and nuclear energy. US oil production is projected to peak around 2025, which will remove the US as the biggest contributor to energy production growth.

The gas production increase in Iran and Qatar and the oil production increase by OPEC keep the Middle East as an important energy source, while the Russian decline in global share of gas production affects its overall global contribution in energy production growth [7].

## 4 USA

The primary energy sources in the USA include petroleum, natural gas, coal, nuclear energy, and renewable. Electricity is a secondary source of energy produced from any of the primary sources 17]. In 2014, the US was largely dependent on fossil fuels, contributing 82% to the total primary energy use in the country. The US energy sector was dominated by nuclear energy, coal, and natural gas, with the electricity supplies and electricity largely consumed by the buildings sector. The transport sector was largely powered by petroleum-based fuels. The improvement in the production of oil and gas is attributed to the use of technology to recover oil and gas from untraditional resources. Lease condensate reserves and proved oil increased by at least 60% between 2010 and 2014, while oil production and natural gas increased by 40% and 25% respectively, in the same period. The annual energy production increase was 18%, with decreased oil imports and lower GHG emissions attributed to the replacement of coal by gas [12] (Fig. 3).

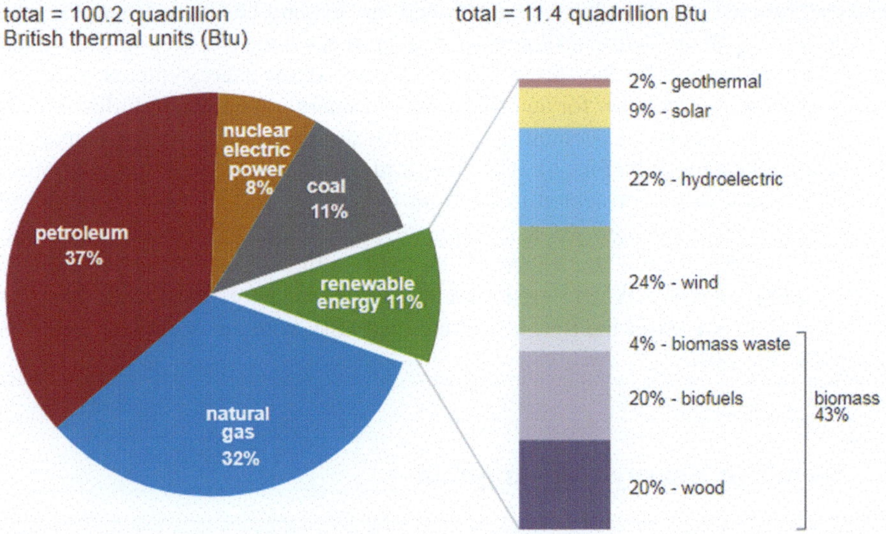


**Fig. 3** US primary energy consumption by energy source, 2019 [13]

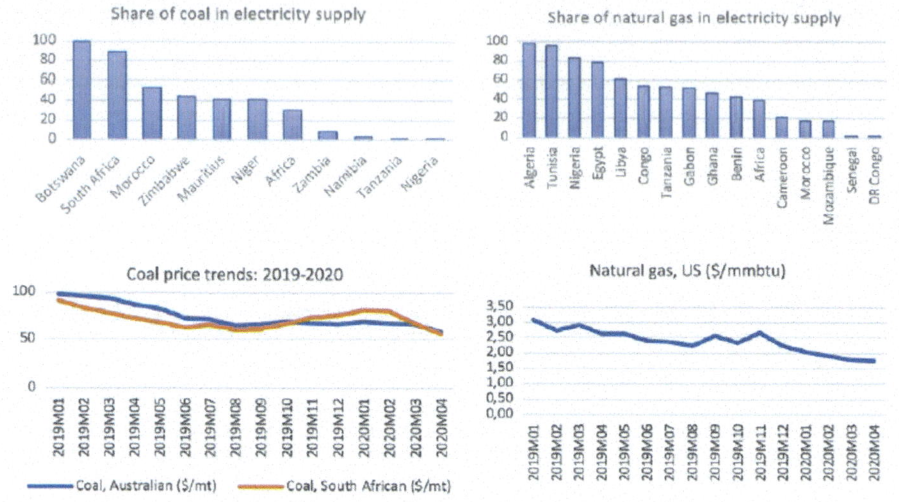


Source: Analysis based on data from the US Energy Information Administration and the International Energy Agency.

**Fig. 4** Share of coal and natural gas in electricity supply and their prices in Africa, from 2019 to 2020 [4]

## 5 Africa and Asia

The energy needs in china have changed, and the demand has tripled over the past two decades, with the industrial energy demand projected to peak around 2025. That has been a motivation for the current transformation to low energy-intensive sectors and construction of policies for an efficient use of energy. As a result, India, Africa, and other Asian countries account for two-thirds of the rise in energy demands for the industrial sector (See Fig. 4). The two-thirds of energy to be used by 2040 in the industrial sector is projected to be from natural gas and electricity. Coal and other Fossil fuel-based energy consumption are projected to decline in China, North America, and the EU, due to cleaner and low carbon fuels [7].

According to pre-COVID-19 estimations, the demand for fossil fuels between 2018 and 2040 was expected to rise, in the global East, with the decline expected in the global West [9].

## 6 Trends in Energy Consumption

The variation in fuel mixes across regions has a crucial impact on the transition of energy. The renewables are predicted to increase mostly in China, EU, US, and India respectively by 2040. Natural gas, nuclear energy, and hydropower are showing growth in shares while coal declines from 60 to 35% between 2017 and 2040, due

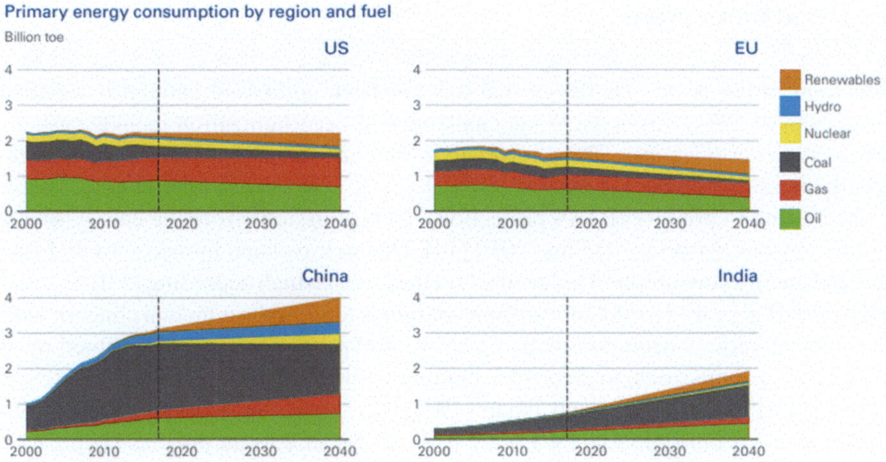


**Fig. 5** Differences in the fuel mix across regions [7]

to sustainability policies in place. China and India, begin with coal intensive mixes in different situations, and both account for the fastest growth in energy demand [4]. There is a balance in the growth of the energy demand in China and the growth of non-fossil fuels. There is a modest decline in the coal share in India, influenced by the growing coal consumption in the Indian power sector. The US and EU begin with diverse mixes of fuel and result in the decline in shares of oil and shares, due to the growing use of natural gas and the renewables in the US [7, 14] (Fig. 5).

## 7 Challenges in the Global Energy Sector

Some challenges in the global energy sector involve declining investments, policies to pace the global warming targets and help the sector meet the projected demand decades to come, capacity and efficiency issues, and $CO_2$ emissions. Most global environmental concerns are linked to energy production [15]. Countries like the US address the challenges of energy security, climate change, and other environmental concerns caused by factors such as public policy, technological readiness, and demand [8]. Clean and renewable energy sources are either limited, costly, or insufficient.

# 8 Decarbonization

Decarbonization is one challenge that has not been addressed enough to balance the growth of $CO_2$ emissions. Some challenges of decarbonization include sustainability and energy supply. The global carbon tax has the potential to accelerate the decarbonization process [16].

According to pre-COVID-19 projections, $CO_2$ emissions from the energy sector will rise by 6% between 2015 and 2050 [14]. Due to travel and human activity bans, reduced energy consumption reduced $CO_2$ emissions, which according to IEA, could fall by 8% this year [9]. Global warming continues to be environmental concern with $CO_2$ atmospheric concentrations being above 400 ppm. USA has contributed most to global $CO_2$ emissions, to date, 25% cumulatively, followed by the EU-28 (22%); China (13%); Russia (6%), and Japan (4%). This means the world is set to fail to reach the target limit of global warming by 2 °C, under current policies [17]. Paced deployment of renewables, improved energy efficiency, and deep electrification can cut energy-related $CO_2$ emissions remarkably by 2050 [1, 18].

Efficiency, cost, and technology readiness are some of the weaknesses of decarbonization that can be addressed by innovations, investments, and R&D. Increasing $CO_2$ concentrations contribute to climate change, which impacts the global energy sector in different ways. In drought-stricken areas, farmers need more energy to power their irrigation systems, which contributes to more emissions since fossil fuels are still dominant in electricity production [19].

# 9 Opportunities

## 9.1 *Electric Cars*

The global energy demand is projected to rise by at least 25% by 2040, with expected increasing demand in developing countries [7, 14] (Fig. 6).

Electric vehicles are predicted to be dominant by 2040, due to climate policies in place and the growth of the field. In 2019, the global sales of electric cars reached 2.1 million, a 40% improvement from the 2018 sales. Electric cars contributed 2.6% to the total car sales globally, and about 1% of worldwide car stock in 2019 [12, 20]. Investments in this sector are can benefit the global energy sector by cutting the burden on fossil fuels and emissions.

## 9.2 *Other Investment Opportunities*

According to IEA, the 2040 global energy demand can be met by the estimated investments worth €1,780 billion in the energy sector, to support the growth and

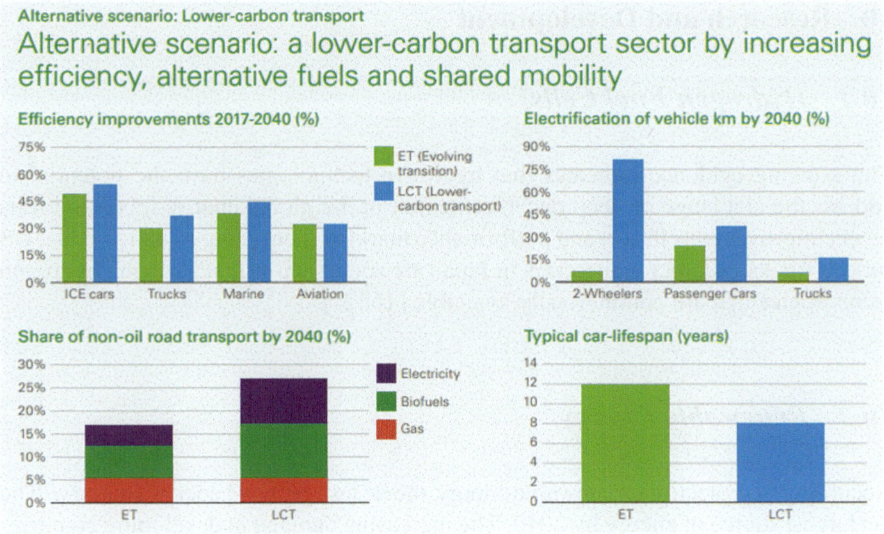


**Fig. 6** Lower carbon transport scenario [6]

development of developing countries [6]. According to some scenarios, investments worth USD 120 trillion will be sufficient for the transition to cleaner energy and to meet the energy demands by 2050 [14]. The transition requires technological readiness, brings new challenges and new opportunities in the energy sector. The improvement in performance and the cost are crucial for the new technologies addressing the challenges in the energy sector.

$CO_2$ emissions have declined since the widespread lockdowns, which is partly due to bans on travel and human activities. However, projections indicate that global warming targets will be missed with the current policies, which open the door for research & development and investments in decarbonization technologies worldwide [17]. Most scenarios project the dominance of renewables by 2040, which can be a confidence booster for investors.

Population growth is one of the drivers of the growth in global energy consumption, with demand in coal, nuclear energy, natural gas, oil, and renewables. Due to the COVID-19 pandemic, a short-term decline in demand for fossil fuels, and a slight growth in renewables is estimated. According to the International Energy Agency (IEA), the estimated decline in primary energy demand is for oil (–9%), coal (–8%), natural gas (–5%), and nuclear energy (–2%) [4, 7, 9].

# 10 Research and Development

## *10.1 Hydrogen Fuel Cells*

The growing evidence indicates that hydrogen technologies have the potential to address the challenge of deep decarbonization in the global energy sector, with its recent improvements in cost and performance marking them economically viable. US forklift trucks and microgeneration in Japan are some sectors that adopting hydrogen technologies that are commercially available [16, 21].

## *10.2 Renewable Energy*

According to projections, renewable energy, the fastest-growing energy source will be the largest source of energy by 2040. The increasing demand in developing countries creates the demand for coal and boosts the penetration of renewable energy [7]. For the first time in history, renewable energy outpaced coal, contributing 23% share, while coal contributed 20% to the total US power generation in April 2019. The contribution of solar and wind to the total US renewable electricity generation was 50%, in early 2019, displacing hydroelectricity. This growth is attributed to the decline in costs, rise in battery storage, and improvement of the capacity of renewable energy sources. The cost of utility-scale solar, onshore wind, offshore wind, and lithium-ion battery storage declined by 18%, 10%, 24%, and 35% respectively. Wind and solar are becoming more competitive with conventional energy sources. The consumption of renewable energy by residential and commercial customers increased by 6% and 5% respectively, with industrial consumption declining by 3% until June 2019, compared to 2018 [4]. The global share of renewable energy was estimated to rise from 14 to 63%, between 2015 and 2050, without consideration of the impact caused by the COVID-19 pandemic in the sector [14].

## *10.3 Solar PV*

There are research projects and prototype development currently taking place to ensure future market growth and efficiency of the solar PVs, such as developing building-integrated photovoltaic solar panels. These solar panels are multifunctional, cost-effective, versatile, flexible for the design size, shape, and color [1, 22–24].

## 11 Conclusion

The global energy sector has many opportunities caused by challenges the sector has been facing including increasing $CO_2$ emissions, decarbonization, efficiency, and the continuously increasing energy demand. Investments in technologies that improve performance, efficiency and cost of the energy resources can help meet the demand in the sector, in the next decades. More investments, research and prototype development should be directed to renewables and solar PV for their continuous growth in various scenarios. Deep electrification of the transport sector, enforcement of climate policies, investments in decarbonization technologies, and research on ways to minimize electricity loses and wastage can help the energy sector to meet the future targets and the demand.

## References


1. https://www.irena.org//media/Files/IRENA/Agency/Publication/2019/Nov/IRENA_Future_of_Solar_PV_2019.pdf.
2. https://www.iea.org/reports/global-energy-review-2020.
3. https://www.worldometers.info/coronavirus/?utm_campaign=homeAdvegas1?.
4. https://www2.deloitte.com/us/en/pages/energy-and-resources/articles/renewable-energy-outlook.html.
5. https://www.iea.org/reports/world-energy-investment-2020/key-findings.
6. https://www.institutmontaigne.org/en/blog/global-energy-demand-how-sustainably-respond-asian-and-african-challenges.
7. https://www.bp.com/content/dam/bp/business-sites/en/global/corporate/pdfs/energy-economics/energy-outlook/bp-energy-outlook-2019.pdf.
8. https://www.energy.gov/sites/prod/files/2015/09/f26/Quadrennial-Technology-Review-2015_0.pdf.
9. https://www.rff.org/publications/reports/global-energy-outlook-2020/.
10. https://www.solarpowereurope.org/wp-content/uploads/2018/09/Global-Market-Outlook-2018-2022.pdf.
11. https://www.edf.org/card/6-ways-cut-big-waste-our-energy-system.
12. https://www.iea.org/reports/global-ev-outlook-2020.
13. https://www.eia.gov/energyexplained/us-energy-facts/.
14. Gielen, D., Boshell, F., Saygin, D., Bazilian, M. D., Wagner, N., & Gorini, R. (2019). The role of renewable energy in the global energy transformation. *Energy Strategy Reviews, 24*, 38–50.
15. Sagar, A. D., & Holdren, J. P. (2002). Assessing the global energy innovation system: Some key issues. *Energy Policy, 30*(6), 465–469.
16. Paradis, E., & Tsatsaronis, G. (2020). Challenges in the decarbonization of the energy sector. *Energy*, 118025.
17. https://ourworldindata.org/co2-and-other-greenhouse-gas-emissions.
18. https://www.fs-unep-centre.org/wp-content/uploads/2020/06/GTR_2020.pdf.
19. http://www.fao.org/3/i9047en/I9047EN.pdf.
20. https://www.ourenergypolicy.org/wpcontent/uploads/2013/09/GlobalEVOutlook_2013.pdf.
21. Staffell, I., Scamman, D., Abad, A. V., Balcombe, P., Dodds, P. E., Ekins, P., et al. (2019). The role of hydrogen and fuel cells in the global energy system. *Energy & Environmental Science, 12*(2), 463–491.
22. https://www.e-education.psu.edu/eme807/node/650.

23. https://www.tralac.org/documents/resources/covid-19/regional/3943-the-impact-of-covid-19-on-africas-energy-sector-res4africa-june-2020/file.html.
24. Felseghi, R. A., Carcadea, E., Raboaca, M. S., Trufin, C. N., & Filote, C. (2019). Hydrogen fuel cell technology for the sustainable future of stationary applications. *Energies, 12*(23), 4593.

# The Impacts of Waste Management and Consideration of Future Policy Direction on Energy from Waste

**Thembinkosi D. Malevu, Zolile Mtumela, Benard S. Mwankemwa, and Richard O. Ocaya**

**Abstract** Managing waste is an interest of many indudtries and it accumulates more attention given the enviromental difficulties and the breakage of viruses. The process of handling waste is managed with the virous options including burning for creating energy. The last and worst option seems to be a burial. Waste burial needs a high volume of land area to be implemented and causes the danger of leakage pollution into air, water, and soil. It also produces less value of energy in comparison with garbage burning. Energy from waste is a broad term that covers much more than waste incineration. It encompasses various waste treatment processes generating energy (e.g., in the form of electricity/or heat or produce a waste-derived fuel), each of which has different environmental impacts and cotribution to circular economy. Waste management is one of the main areas where further improvements are needed and within reach: increasing waste prevention, reuse and recycling are key objectives, both of the action plan and of the legislative package on waste. In seeking to solve the problem of waste disposal and producing energy from waste is one of the most challenging problems of the current world. To manage waste effectively, public–private partnership should be encouraged to jointly address waste management problems. The legislation, policies, governmental rules, and new technologies should be considered worldwide to raise the situation of waste to energy conversion in the market of renewable energies.

T. D. Malevu (✉)
Department of Physical and Chemical Sciences, North-West University (Mafikeng Campus), Mahikeng, South Africa
e-mail: Thembinkosi.malevu@nwu.ac.za; Malevu.td@gmail.com

Z. Mtumela
Department of Chemical and Physical Science, Walter Sisulu University, Mthatha, South Africa

B. S. Mwankemwa
Department of Physics, School of Physical Sciences, College of Natural and Mathematical Sciences, University of Dodoma, Dodoma, Tanzania

R. O. Ocaya
University of the Free State, Qwaqwa Campus, Bloemfontein, South Africa

 
M. Tshwafo Elias (ed.), *Waste to Value Innovations*,
https://doi.org/10.1007/978-981-96-2782-0_9

**Keywords** Waste management · Waste to energy · Legislations · Recycling methods · Pyrolysis

## 1 Introduction

In South Africa and African continent at large, there are lot of waste which are generated on daily basis. These waste materials have proven to have a negative impact on the environment by causing air, water pollution, and human health [1]. To address the issues, a proper waste management program has to be implemented. One of the options to manage waste is changing waste into useful energy. Waste management is a cost-effective process in which South Africa and African continent Authorities incurred. Global municipal waste generation nowadays is appromately 1.3 billion tons per year which is expected to increase to 2.2 billion tons by year 2025 [2]. A number of factors such as urbanization, economic development, lack of infrastructure and population growth are considered as a main cause of gradual increase of waste in South Africa [3, 4]. For instance, South Africa's re-integration into the global economy and the Southern African political arena necessitates an improved pollution and waste management system.

The recent study survey conducted by the South African Waste Sector shows that landfilling remains the main technology option for managing waste in South African, with 90% of all waste generated disposed [5]. The results show a high level of waste innovation activity, with the private sector showing greater innovation activity than the public sector. The estimation of 51.9% of private sector compared to 41.2% of municipality sectors indicated they had introduced new product innovations during the past years. However, given the current constraints facing the South African waste sector, creating an enabling environment for waste innovation is imperative to changing the sectors dependency on landfilling.

An economic development, a growing population and increasing rates of urbanization in Southren Africa have resulted in increased waste generation which requires establishing and implementing effective waste management policies and programmes [6]. A number of issues continue to be challenges for effective waste management. These include ineffective data collection systems and lack of compliance and enforcement capacity, lack of education and awareness amongst stakeholders within the waste sector, operational costs for management of waste, support for waste reduction at local government level, availability of suitable land for waste disposal, lack of structured incentives for reduction, and recycling and/or reuse of waste [7, 8].

Southern African political arena necessitates an improved pollution and waste management system. This has also seen South Africa's economic and industrial policy turned towards export promotion as a pillar of South Africa's development. Therefore, South Africa has a growing obligation to meet international commitments and to be a globally responsible nation. The government therefore promotes an integrated approach to pollution and waste management as a key factor in achieving sustainable development.

The integrated pollution and waste management policy is driven by a vision of environmentally sustainable economic development [9]. This vision promotes a clean, healthy environment, and a strong, stable economy. By preventing, minimizing, controlling and mitigating pollution and waste, the environment is protected from degradation by enhancing sustainable development. The constitution of South Africa (Act 108 of 1996) [10] established the Bill of Rights that ensures that everyone has the right to an environment that is not harmful to their health and well-being. Legislative and other measures should be used to ensure that the environment is conserved and protected for future generations. The policy framework to move waste up the waste management hierarchy away from landfilling towards waste reuse, recycling and recovery has been provided by government through the promulgation of the National Environmental Management: Waste Act (NEM:WA) (RSA 2008) [11]. However, limited understanding of the main waste flows and national waste balance because the submission of waste data is not obligatory, and where data is available, it is often unreliable and contradictory; The waste is catagorised into 3 definitions as defined in NEM:WA [11] are presented below.

'general waste' means waste that does not pose an immediate hazard or threat to health or the environment, and includes: Domestic waste; Building and demolition waste; Business waste; and, Inert waste.

'hazardous waste' means any waste that contains organic or inorganic elements or compounds that may, owing to the inherent physical, chemical or toxicological characteristics of that waste, have a detrimental impact on health and the environment.

'building and demolition waste' means waste, excluding hazardous waste, produced during the construction, alteration, repair or demolition of any structure, and includes rubble, earth, rock and wood displaced during that construction, alteration, repair or demolition.

## 2 Energy from Waste Technology Applications

The energy from waste can be directly derived by converting waste into useful energy applications in the form of biogas, syngas or heat. These conversion processes can be done through physical, thermal and biological methods as shown in Fig. 1.

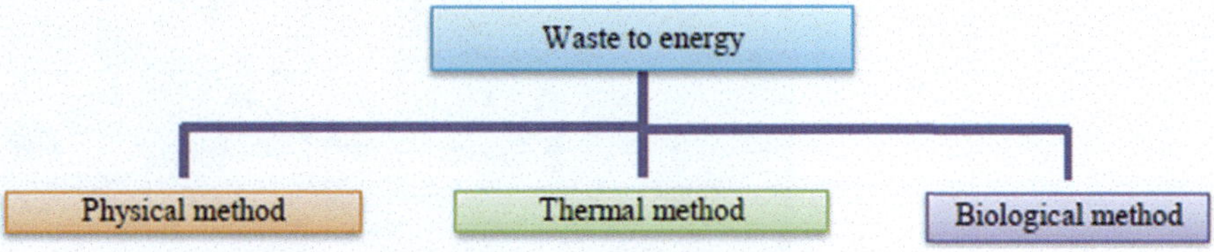


**Fig. 1** Schematic diagram of waste to energy management methods [12]

### Thermal Method (Incineration)

Thermal method involves the combustion of organic materials into incinerator bottom ash, gases, particulates, and heat that can be used to generate electricity [13]. It is both a landfill reduction method, reducing the volume of waste by 95–96%, and energy from waste technology. The heat produced during incineration can be used to generate steam which can then be used to drive a turbine in order to produce electricity. The typical range of net electrical energy that can be produced is about 500–600 kWh of per tonne of waste incinerated. Thus, the incineration of about 2,200 tonnes per day of waste could produce about 50 MW of electrical power. Electricity production is thus around 0.5–0.7 MWh/t of municipal solid waste [14]. Incineration reduces mass and volume of landfill, lightening the load of landfill management in cities. There has been concern around the health ramifications of incineration, but significant advances in emission control have occurred and strict regulations have been initiated concerning dioxin and furan emissions (both of which are highly toxic substances). By diverting municipal solid waste, incineration avoids the release of methane into the atmosphere. In addition to methane, for every tonne of municipal solid waste that gets incinerated, approximately one tonne of carbon dioxide ($CO_2$) is prevented from being released into the atmosphere.

Medical waste incineration is the most common application in South Africa, but there are also some industrial waste incinerators, especailly those involving hazardous waste. In fact, there are designed mobile incinerators to curb high volume of medical waste as shown in Fig. 2.

The biggest challenges for waste-to-energy mass incineration currently in countries like South Africa lie in the high capital costs and air emission control requirements. Incineration produces particles containing toxic metals, dioxins, and furans that are so small that they can potentially evade pollution control devices. Incineration also produces a highly toxic fly ash that must be safely disposed of; leading to transportation and residential health concerns [15]. Gasification and pyrolysis on the other hand are cleaner processes and do not pose toxicity threats, but remain "third

**Fig. 2** An example of a low capacity, mobile incinerator

generation" technologies. Although the fuel (i.e. the waste) in energy from waste incineration plant may be free, and the plant may save hugely on landfilling costs, the high capital costs of plants means that incineration is still prohibitively costly as a household waste treatment approach in South Africa.

**Gasification/Pyrolysis**

Like all organic materials including biomass, wood and plastic waste, can be converted into a gas mixture of carbon monoxide, hydrogen and carbon dioxide by gasification and pyrolysis. This is achieved by reacting the material at high temperatures (>700 °C), without combustion, with a controlled amount of oxygen and/or steam. Unlike incineration, gasification does not produce energy from waste through direct combustion. Waste, steam, and oxygen are fed into a gasifier where heat and pressure break apart the chemical bonds of the waste to form synthesis gas (syngas). It allows the breakdown of hydrocarbons into the gaseous mixture by carefully controlling the amount of oxygen available. Syngas may be used directly in internal combustion engines or to make products that substitute for natural gas, chemicals, fertilisers, transportation fuels and hydrogen. Pollutants are removed from syngas before it is combusted, so that it does not produce the high levels of emissions associated with other combustion technologies. Like gasification, pyrolysis also turns waste into energy by heating under controlled conditions, but involves thermal degradation in the complete absence of air. Pyrolysis typically occurs under pressure and at operating temperatures above 430 ºC.

Gasification and pyrolysis are extremely efficient ways of using biomass to produce energy, both being more efficient than incineration. They are flexible technologies where existing gas-fuelled devices (ovens, furnaces, boilers, etc.) can be retrofitted with gasifiers and syngas can directly replace fossil fuels. It is able to generate energy which is cheaper and more efficient than the steam process used in incineration. Municipal solid waste can be reduced by as much as 75% through this process, reducing to the same degree the amount of potential emissions the waste would have created in a landfill. The process of sorting and preparing the solid waste (autoclaving) for pyrolysis is well established and the technological expertise is available in South Africa. These technologies are cleaner than incineration and do not pose toxicity threats. However, the technology is still relatively new (or "third generation") with limited plants in operation around the world (although anticipated to grow substantially into the future).

## 3 Waste Management Approach: Energy Recovery in the Context of the Waste Hierarchy

In a perfect world, all waste would be avoided and the hierarchy would be unnecessary. However, in reality for a range of social, economic, practical and technological reasons, this never happen. Where waste does exist it is usually best to re-use it if possible, and if not, to recycle it. Waste management is one of the main areas where further improvements are needed and within reach: increasing waste prevention, reuse and recycling are key objectives both of the action plan and of the legislative package on waste. In South Africa, the approach to waste management is structured around the hierarchy. Figure 3 shows a South African waste hierarchy ranging from waste materials into primary and secondary products. The hierarchy is applied in the planning system through government waste management policy. South African waste legislation is influenced and informed by the key elements of the waste hierarchy, which dictates the overall strategic approach for waste management. The waste hierarchy is also clearly visible in the NWMS for South Africa.

This waste hierarchy approach towards waste management emphasizes the following key elements [1, 8]:

- **Prevent and Reduction**: Products and materials must be designed in a manner that minimizes their waste components or that reduces the natural material quantities

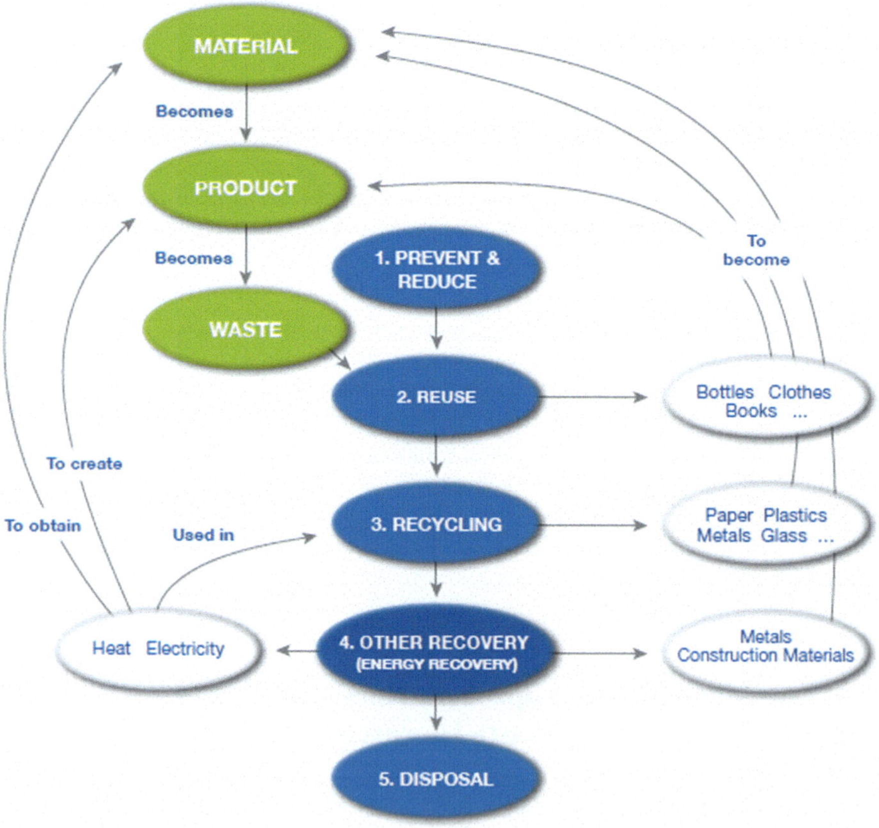


**Fig. 3** Waste management hierarchy [8]

used and potential toxicity of waste generated during the production, and after use;

- **Reuse**: Materials can be used for similar or different purposes without changing form or properties. This approach seeks to re-use a product when it reaches the end of its life span. In this way, it becomes input for new products and materials;
- **Recycling**: This involves separating materials from the waste stream and processing them as products or raw materials. The first elements of the waste management hierarchy are the foundation of the cradle-to-cradle waste management approach;
- **Recovery**: Reclaiming particular components or materials or using the waste as a fuel;
- **Treatment and disposal**: This is a 'last resort' within the waste hierarchy. Treatment refers to any process that is designed to minimize the environmental impact of waste by changing the physical properties of waste or separating out and destroying toxic components of waste. Disposal refers specifically to the depositing or burial of waste onto, or into land; and,
- **Legal**: Processing, treatment, and disposal of waste must take place in accordance with the principles of environmental justice and equitable access to environmental services.

## 4 Waste Management Challenges

Waste management in Southren Africa has in the past been uncoordinated and poorly funded. According to Nahman and Godfrey [15] and Park et al. [9] key issues include inadequate waste collection services for a large portion of the population, illegal dumping, unlicensed waste management activities (including unpermitted disposal facilities), a lack of airspace at permitted landfills, insufficient waste minimization and recycling initiatives, a lack of waste information, lack of regulation and enforcement of legislation, and, indeed, limited waste- related legislation in the first place. In response, the National Waste Management Strategy (NWMS), Department of Environmental Affairs and Tourism [DEAT], emphasizes the need for integrated waste management [11], which implies coordination of functions within the waste management hierarchy. In particular, the diversion of waste from landfill through waste minimization and recycling is a national policy objective under the White Paper on Integrated Pollution and Waste Management (Department of Environmental Affairs and Tourism [10], the NWMS and the Waste Act, which recognize the importance of moving waste management up the waste hierarchy. In addition, to deal with the issue of insufficient funding, the NWMS invokes the Polluter Pays Principle (PPP). In the context of solid waste management, the PPP implies that all waste generators, including households and companies, are responsible for paying the costs associated with the waste they generate. These include not only the direct costs associated with the safe collection, treatment and disposal of waste; but also the external costs of waste generation and disposal, such as health and environmental damages [11].

## 5 The Environmental Legislative Framework in Southren Africa

Hazardous wastes, in particular, require more stringent regulatory and technical controls due to their toxicity, persistence, mobility and flammability. There is increasing public concern about the numerous problems and potentially dangerous situations associated with hazardous waste management in general and disposal practices in particular [12].

Countries like South Africa has introduced a range of legislative measures aimed at improving the quality of the environment. The effective regulation of hazardous wastes requires sufficient compliance and enforcement capacity on the part of Department of Environmental Affairs. Waste in South Africa is currently governed by means of a number of pieces of legislation, including:

- The South African constitution Act 108 of 1996
- Hazardous Substance Act 5 of 1973
- Environmental Conservation Act 73 of 1989
- National Water Act 36 of 1998
- National Environmental Management Act 107 of 1998
- Minerals and Petroleum Resources Development Act 28 of 2002
- Air Quality Act 39 of 2004
- National Environmental Management: Waste Act 59 of 2008.

To elaborate on these legislation the Environmental Management Policy for South Africa sets a number of objectives for integrated pollution control and waste management system.

The objectives include:

- Promoting cleaner production and establishing mechanisms to ensure continuous improvements in best practices in all areas of environmental management.
- Preventing or reducing and managing pollution of any part of the environment due to all forms of human activity, and in particular from radioactive, toxic and other hazardous substances.
- Setting targets to minimize waste generation and pollution at source and promoting a hierarchy of waste management practices, namely reduction of waste at source, reuse and recycling with safe disposal as the last resort.
- Regulating and monitoring waste production, enforce waste control measures, and coordinating administration of integrated pollution and waste management through a single government department.
- Setting up information systems on chemical hazards and toxic releases and ensuring the introduction of a system to track the transport of hazardous materials.

The South African waste management principles aim:

- To secure the conservation of nature and resources, waste generation must be minimized and avoided where possible (prevention principle).

- To secure a reduction in the impacts from waste on human health and environment, especially to reduce the hazardous substances in the waste through precautionary principle.
- To make sure that those who generate waste or contaminate the environment should pay the full costs of their actions through the principle of pollute pays and producer responsibility.

In relation to the mining waste, the strategic focus in terms of waste hierarchy is on ensuring the treatment and safe disposal of mining waste. However, opportunities for reuse of mining waste need to be fully exploited. The overall goal with regard to regulating waste invariably is to minimize health and environmental impacts with the concurrent optimization of economic and social impacts on society.

## 6 Energy Outputs

Energy from agriculture waste could either convert the waste to energy in the form of electricity and heat, or convert it to 'transportable' energy-rich commodities such as fuel, substitute (or synthetic) natural gas (SNG) for injection into the grid, hydrogen and simple chemicals that could be used to make new materials. As noted, the greater the *overall* efficiency with which the waste is turned into useful energy, that is 'exportable' energy net of any energy required by the process to create it, the better the environmental performance of energy from waste. Indirect generation of electric energy, such as burning waste to create steam for pressure in a turbine is generally less efficient than direct energy generation. However, it will often require energy, 'parasitic load', to create the fuel for direct generation so the balance in terms of overall efficiency can be less clear and needs to be considered in determining the relative value and environmental benefit of the outputs. The level of financial risk associated with different outputs also varies as some technologies are less mature than others.

## 7 Electricity

The conventional way to generate electricity from waste is through direct combustion, with the heat used to produce steam to drive a turbine. Combustion is the process of burning the waste directly to produce heat which can be used for useful application for example space heating and boiling water into steam for power generation. This is indirect generation with an overall efficiency of 15–27%. Modern plants tend to be at the top end of this range. Steam generation from gasification is no more efficient than from incineration and due to lower operating temperatures, steam pressure and parasitic loads (i.e., energy required to run the plant) the overall process may be less efficient than conventional incineration. Processes such as gasification and pyrolysis produce a combustible synthesis gas (often abbreviated to syngas) which can either be used to raise steam as described above or more innovatively can be cleaned up for direct generation in a gas turbine or engine. Gas turbines or engines can provide greater efficiency in the energy generation step than steam turbines and thus there is potential for gains in overall efficiency if parasitic load, especially from the syngas clean-up, can be minimized.

## 8 Institutional Drivers

- A policy and regulatory environment that does not actively promote the waste management hierarchy. This has limited the economic potential of the waste management sector, which has an estimated turnover of approximately R10 billion per annum. Both waste collection and the recycling industry make meaningful contributions to job creation and GDP, and they can expand further;
- Growing pressure on outdated waste management infrastructure, with declining levels of capital investment and maintenance;
- Waste management suffers from a pervasive under- pricing, which means that the costs of waste management are not fully appreciated by consumers and industry, and waste disposal is preferred over other options; and,
- Too few adequate, compliant landfills and hazardous waste management facilities, which hinders the safe disposal of all waste streams. Although estimates put the number of waste handling facilities at more than 2,000, a significant number of these are unlicensed.

## 9 Waste Disposal and Treatment in South Africa

Waste management services rely heavily on landfills for the disposal of waste, which account for the majority of licensed waste facilities. Over 90% of all South Africa's waste is disposed of at landfill sites [13]. This is despite the existence of a range of alternative management options, including waste recycling facilities. Table 1 illustrates the number of licensed waste management facilities in the country.

**Table 1** Waste management facilities license status in South Africa [13]

| | | | |
|---|---|---|---|
| General waste landfill site | 1.203 | 432 | 56.4 |
| Hazardous waste landfill site | 77 | 86 | 46.8 |
| Health care risk waste storage facility | 25 | 25 | 100.0 |
| Recycling facilities | 9 | 44 | 77.8 |
| Transfer stations | 35 | 88 | 65.7 |
| Total | 1.336 | 675 | 56.4 |

Waste disposal by landfill remains the most dominant method of disposal in South Africa, and the reliance on which, coupled with the relative low pricing for landfilling, has limited the incentive to devise alternative methods of dealing with waste. Furthermore, there is an urgent need to address the licensing status of landfill sites; compliance to license conditions must be enforced. Many landfill sites that are operating which are non-compliant with license conditions. This is resulting in poor levels of operation and negative impacts on the environment [16–18]. It is critical that all waste management facilities are licenced in order to avoid potential negative environmental impacts, as it is through the licensing process that any significant impacts are identified, and mitigation actions prescribed.

## 10 Waste Management Trends in South Africa

Environmental issues linked to the increased use of landfill sites as the dominant means of waste disposal and treatment have in recent years become a matter of public concern and environmental awareness is growing as a result of the pressure to change this behaviour. The third National Waste Information Baseline Report estimated [14] that South Africa generated approximately 108 million tonnes of waste in 2011, of which 98 million tonnes was disposed of at landfill. It was estimated that 59 million tonnes was general waste, 48 million tonnes currently unclassified waste and the remaining one million tonnes hazardous waste. In the order of ten per cent of all waste generated in South Africa was recycled in 2011.

## 11 Municipal Solid Waste

Municipal waste is not the largest waste category by volume in South Africa (the largest waste category is industrial and mining waste), but it is the most significant in terms of public financing and the impact that it has on the day-to-day lives of ordinary citizens [15]. The classification of general waste as per the Waste Classification Regulations are presented in Table 2.

According to the NWIBR [18], South Africa generated 59 million tonnes of general wastein 2011. An estimated 5.9 million tonnes of general waste was recycled (ten per cent) with the remaining 53.5 million tonnes of general waste being landfilled. Municipal (GW01) and commercial and industrial waste (GW10), generated within municipalities represents a total of 20,157,335 tonnes when the recyclables (reported under GW30, GW50, GW51, GW52, GW53 and GW54) are included. The percentage contribution of each waste stream to the composition of general waste is illustrated in Fig. 4. Non-recyclable municipal waste contributes 35 percent (by weight) of the overall general waste, construction and demolition waste, 20%, followed by metals (13%), organic waste 13% and mainline recyclables (including paper, plastics, glass and tyres (19%) [20].

**Table 2** General waste categories [19]

| | |
|---|---|
| GW01 | Municipal waste |
| GW10 | Commercial and industrial waste |
| GW13 | Brine |
| GW14 | Fly ash and dust from miscellaneous filter sources |
| GW15 | Bottom ash |
| GW16 | Slag |
| GW17 | Mineral waste |
| GW18 | Waste of Electric and Electronic Equipment (WEEE) |
| GW20 | Organic waste |
| GW21 | Sewage sludge |
| GW30 | Construction and demolition waste |
| GW50 | Paper |
| GW51 | Plastic |
| GW52 | Glass |
| GW53 | Metals |
| GW54 | Tyres |
| GW99 | Other |

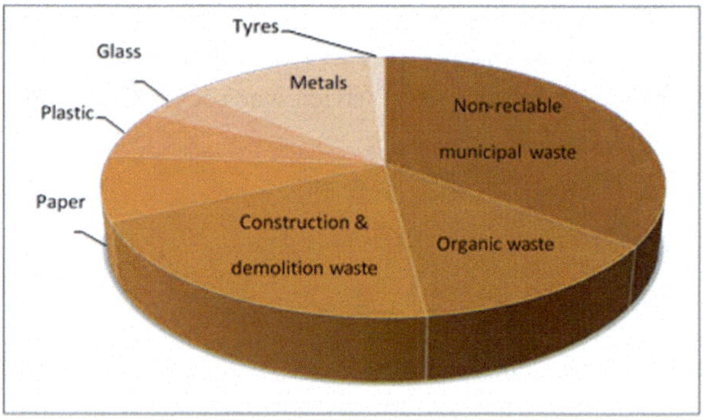


**Fig. 4** Waste composition for general waste in 2011 shown as percentage by mass

## 12 Economics and Financing of Waste Management

Various research investigations have been attempted on waste administration in South Africa, however most have generally centered around operational administration, ecological viewpoints and the authoritative system. A little committent has been done to adress the model of financing and policies. According to Resource and wase advisory group (RWA) [19], advanced waste treatment is generally more expensive than landfilling when considering the financial costs of the service only. However, when considering the wider economic costs and benefits from a societal standpoint, the advanced treatment options become more favourable than just landfill. Currently, municipal councils, which are responsible for budget allocations, do not recognise waste management as a priority service. As a result, waste management budgets tend to be relatively low compared to those of other services [21]. Therefore, when planning waste management improvements, municipalities need to concider investment costs and funders. As an informal system is still playing a significant role in waste accumulation, the local governments should consider organizing and managing this informal system so that it can be better regulated by the municipal institutions. Not only would this improve the efficiency and cleanliness of urban solid waste collection, it would also provide job opportunities for these informal waste. In many occasion, recycling and recovery is usually conducted by the informal sector on all levels of the waste management stream. Such work is done in a very labour-intensive and unsafe way, and for very low incomes. The situation in industrialised countries is very different, since resource recovery is undertaken by the formal sector, driven by law and a general public concern for the environment, and often at considerable expense.

## 13 Future Policy Path

Waste management is changing to be much less about how waste is removed from sources and more about introcution of cricular economy models. That could be witnessed in the novel energy generation by means of renewables and polymers in general.

The South African government sees a long-term role for energy from waste both as a waste management tool and as a source of cheap energy. Energy from waste is in a unique position to fulfil a range of objectives across a number of South Africa government departments. This can contribute to growth in the waste and energy sectors as well as the construction sector through infrastructure development. There are wider societal and environmental benefits associated with energy generation and use that will drive energy policy and impact on energy from waste. Energy from waste could be subject to policy developement to address these drivers. Examples might include renewable energy targets and how these are used amongst electricity, heat and transport fuels. Energy from waste in particular has the potential to deliver low carbon energy in a cost effective way and as a non-intermittent source helps provide energy security. There is therefore a range of policies which waste management and energy generation companies have to navigate. With the average contract life time of an energy from waste plant being 25 years, potential investors need to be mindful of South Africa government's vision for the future. Issues of concern and lessons that have been identified and require significant attention in development of Integrated Waste Management Planning include recognition and support of community waste management and servicing as well as trans-boundary effects on environmental quality. There is a need to enhance private and public sector partnerships in waste management initiatives.

## 14 The Principles Underpinning the Policy

There are four key principles that underpin current thinking on energy from waste and which are expected to remain critical to the development of a sustainable policy into the future. These principles are outlined as follows:

- Energy from waste must support the management of waste in line with the waste hierarchy.
- Energy from waste should seek to reduce or mitigate the environmental impacts of waste management and then seek to maximize the benefits of energy generation.
- A support for energy from waste should provide value for money and make a cost effective contribution to environmental objectives in the context of overall waste management and energy goals.
- South Africa government must remain technology neutral except where there is a clear market failure preventing a technology competing.

South Africa government must encourage developers to consider these principles as a key part of the decision making process around future development of new projects and operation of existing plant. This means that from a sector viewpoint infrastructure proposals, technologies and services that are aligned with these principles should be on a much firmer footing and more robust to future policy than those which are not.

## 15 Conclusion

The amount of waste must be decreased, otherwise, recycling or re-use are the other options. Another option is burning or energy recycling. If applying these options were not possible, the last and worst option was burial. Waste burial needs a high volume of land area to be implemented and causes the danger of leakage pollution into air, water, and soil; it also produces less value of energy in comparison with garbage burning. In general, it is supposed that burning has less lateral effects, especially in the production of energy from waste. In seeking to solve the problem of waste disposal and producing energy from waste is one of the most challenging problems of the current world. To manage waste effectively, public–private partnerships should be encouraged to jointly address waste management problems. Since waste materials have been accepted as a clean source of energy, policies, governmental rules, and new technologies should be considered worldwide to raise the situation of waste to energy conversion in the market of renewable energies. In addition, South Africa government must encourage developers to consider principles mentioned here as a key part of the decision-making process around the future development of new projects and operation of existing plant. This means that from some sector viewpoint infrastructure proposals, technologies and services that are aligned with these principles should be on a much firmer footing and more robust to future policy than those which are not.

**Declaration of Conflicting Interests** The author(s) declared no potential conflicts of interest with respect to the research, authorship, and publication of this article.

**Funding** The author(s) received no financial support for the research, authorship, and publication of this article.

## References

1. Scarlat, N., Motola, V., Dallemand, J. F., Monforti-Ferrario, F., & Mofor, L. (2015). Evaluation of energy potential of municipal solid waste from African urban areas. *Renewable and Sustainable Energy Reviews, 50,* 1269–1286.
2. Ouda, O. K., & Raza, S. A. (2014, May). Waste-to-energy: Solution for municipal solid waste challenges-global perspective. In *2014 International Symposium on Technology Management and Emerging Technologies* (pp. 270–274). IEEE.
3. Diamond, J. (2002). Evolution, consequences and future of plant and animal domestication. *Nature, 418*(6898), 700.
4. Boadi, K. O., & Kuitunen, M. (2002). Urban waste pollution in the Korle lagoon, Accra, Ghana. *Environmentalist, 22*(4), 301–309.
5. DEA (Department of Environmental Affairs). (2009a). *National waste quantification and waste information system.* Department of Environmental Affairs, Pretoria.
6. DEA (Department of Environmental Affairs). (2009b). *National waste information regulations in terms of the national environmental management: Waste Act (No. 59 of 2008) notice 718 of 2010.* Department of Environmental Affairs, Pretoria.
7. Kitazawa, S., & Sarkis, J. (2000). The relationship between ISO 14001 and continuous source reduction programs. *International Journal of Operations & Production Management, 20*(2), 225–248.
8. South Africa. (1994). *Constitution of the Republic of South Africa, 1993* (Vol. 8). Creda Press for the Government Printer, Pretoria.
9. Shelton, D. (Ed.). (2003). *Commitment and compliance: The role of non-binding norms in the international legal system.* Oxford University Press on Demand.
10. Park, H. S., Rene, E. R., Choi, S. M., & Chiu, A. S. (2008). Strategies for sustainable development of industrial park in Ulsan, South Korea—from spontaneous evolution to systematic expansion of industrial symbiosis. *Journal of Environmental Management, 87*(1), 1–13.
11. Royal Haskoning, D. H. V. (2014). Municipal solid waste diversion and beneficiation opportunities at Nelson Mandela bay metro municipality–feasibility study final report.
12. Pandey, B. K., Vyas, S., Pandey, M., & Gaur, A. (2016). Municipal solid waste to energy conversion methodology as physical, thermal, and biological methods. *Current Science Perspectives, 2,* 39–46.
13. Lees, Jr., J. N., Svedala Industries Inc. (1994). Method of capturing and fixing volatile metal and metal oxides in an incineration process. U.S. Patent 5,333,558.
14. European Suppliers of Waste to Energy Technology. (2012). *Everything you always wanted to know about waste to energy.*
15. Ferreira, C., Ribeiro, A., & Ottosen, L. (2003). Possible applications for municipal solid waste fly ash. *Journal of Hazardous Materials, 96*(2–3), 201–216.
16. Nahman, A., & Godfrey, L. (2010). Economic instruments for solid waste management in South Africa: Opportunities and constraints. *Resources, Conservation and Recycling, 54*(8), 521–531.
17. DEA (Department of Environmental Affairs). (2011). *Draft municipality waste sector plan.* Department of Environmental Affairs, Pretoria.
18. DEA (Department of Environmental Affairs). (2012a). *National waste management strategy.* Department of Environmental Affairs, Pretoria.

19. DEA (Department of Environmental Affairs). (2012b). *National waste information baseline report*. Department of Environmental Affairs, Pretoria.
20. DEAT (Department of Environmental Affairs and Tourism). (2000). *White paper on integrated pollution and waste management for South Africa: A policy on pollution prevention, waste minimisation, impact management and remediation*. Department of Environmental Affairs and Tourism, Pretoria.
21. Zurbrugg, C. (2003). *Solid waste management in developing countries* (Vol. 5). SWM introductory text on www.sanicon.net.

# Food Waste and Biofuel

**Nhlapo P. Lefu and Motaung Tshwafo**

**Abstract** Apart from COVID-19 pandemic, the world is confronted with many challenges which include inter alia environmental and economic challenges, increasing population and depletion of natural resources. In this paper we will focus a specific focus on food waste, its impact on the environment and its conversion to useful products i.e., biofuels. According to the Food and Agricultural Organization (FAO), one third of food produced globally for human consumption is lost along the food supply chain. Food waste is becoming an increasingly significant global issue due to its economic, social and environmental implications as waste is being created across the entire food supply chain, from farm to fork. From environmental point of view, food waste relates to climate change since food waste is still being landfilled or incinerated in many countries together with other combustible municipal wastes for possible recovery of energy heat or other forms of energy. The challenge here is the residual ash that is disposed of in landfills, the resultant air pollution from the incineration process as well as the cost associated with waste conversion technique. However, owing to their organics- and nutrients-rich composition food waste is viewed as a useful resource to produce biodiesel, bioethanol and biogas (i.e., biomethane and biohydrogen) using processes such as transesterification, alcoholic fermentation and biomass anaerobic, respectively. Biodiesel obtained from food waste oils and used vegetable cooking can be used as an alternative to petroleum fuels without modifying the engine whilst the biogas is utilized in many countries to generate electricity thereby significantly reducing the dependency of economy on fossil fuels, reducing the emission of greenhouse gases. In some countries gasohol (mixture of gasoline and bioethanol) is used to fuel vehicles.

N. P. Lefu
Sasol Operations, P.O. Box 1, Sasolburg, Gauteng, South Africa

M. Tshwafo (✉)
Department of Chemistry, Sefako Makgatho Health Sciences University, P.O. Box 94, Medunsa 0204, South Africa
e-mail: tshwafo.motaung@smu.ac.za; motaungte@live.com

 
M. Tshwafo Elias (ed.), *Waste to Value Innovations*,
https://doi.org/10.1007/978-981-96-2782-0_10

# 1 Introduction

The by-product, excess products of various food processing industries, commonly referred to as food waste, has economic, environmental and social implications and impacts in the world [1–3]. Though there is no unified agreement on the definition of food waste [4], the term food waste generally refers to waste and residue streams throughout the food value chain from primary production, processing, distribution to end use. Here the primary production will include, inter alia the cereal straw and other wastes and residue streams generated in agriculture and farming and processing covers the waste streams from the food industry whilst the distribution involves waste streams retail and wholesale. End use simply refers to food service and households generating various food waste streams. Food is lost or wasted throughout the entire food supply chain viz. agricultural production and post-harvesting and storage, processing, distribution to final household consumption.

According to researchers [1, 2], most of food wastes are generated in developed countries post-farmgate where consumers are buying excessive food and throwing away unconsumed stuff. In 2014, the Hong Kong Environment Bureau of 9000 tonnes of municipal waste that is thrown away every day at landfills, roughly 90% of putrescibles are food wastes [1].

It was also reported [2, 3] that approximately 1.3 billion tonnes of foods waste were thrown away back in 2016 and this equaled to one-third of food produced for human consumption globally. In 2011, Hong Kongers was found to throw away Ca. 3,600 tonnes of food waste every day. Of 3,600 two-third was obtained from the household whereas one-third came from commercial and industrial resources. In the United States alone in 2012 [5], consumers were reported to waste 10 times more food per capita than those in Southeast Asia while UK estimated 10 million tonnes of food waste arising post farmgate each year [3]. In South Africa, food waste generation was estimated to be 9 million tons per annum [6] whilst between 2006 and 2012 Singapore reached about 703,200 tons from 542,720 tons of food waste production [7]. It must, however, be noted that in many countries food waste is currently landfilled or incinerated together with other combustible municipal waste for possible recovery of energy. The absurd methodology obviously adds pain to the environment in terms of occupying the productive space and air pollution.

To curb the skyrocketing food waste landfills, the governmental bodies and political institutions both in European Union [8, 9] and in the United States [10] developed the food waste recovery hierarchy chart shown in Fig. 1. According to the chart the primary focus in handling food waste should be sheer prevention of food waste at all costs. In an event impossible to prevent, then the preferred measure is to avail the surplus for human consumption, as shown by tiers 1 and 2. This is, however, a challenge as the food and Agricultural Organization (FAO) reported that one third of food produced globally for human consumption is lost along the food supply chain [11] probably due to the fact that a large fraction of wasted food is perishable and not always edible [2]. If food surplus cannot be donated, the next tier suggests the food waste should be used as animals feed with the least preferred relating to recycling

**Fig. 1** Food recovery hierarchy [3, 12]

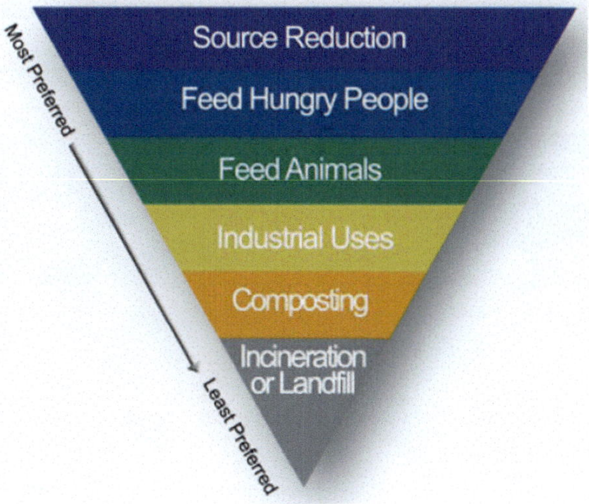


food waste into compost or converting it into biogas or biodiesel. The least preferred option is the disposal of food waste into landfills or incineration.

The increasing political and scientific consensus regarding the need to reduce global food waste [4], the intensified need to phase out fossil fuels [11], political issues connected with important oil-producing countries and the emission of greenhouse gases and climate change have moved the focus of many governments to explore alternatives of using renewable energy sources [13]. Moreover, the United Nations' Sustainable Development Goal 12.3 ambitiously announced its set the target of 2030 to halve per capita global food waste at the retail and consumer levels and reduce food losses along production and supply chains, including post-harvest losses [4, 14].

## 2 Forecasts of Food Waste

The share of global food that is thrown away has risen significantly over the years. Reduction and effective management of food waste have become a pivotal challenge for industries and communities alike. The rising food losses and waste are expected to have detrimental environmental and economic impacts on society. Due to the increasing world population, food wastage is becoming a serious issue around the world. The expanding growth in the amount of food waste will have serious adverse effects on society, such as health risks, environmental pollution and scarcity of landfills. There is a consistently growing need to create appropriate, environmentally sustainable solutions to manage the increasing amount of food waste. Among the solutions, the anaerobic digestion process is one of the most promising and environmentally friendly solutions which can help in food waste management, nutrient

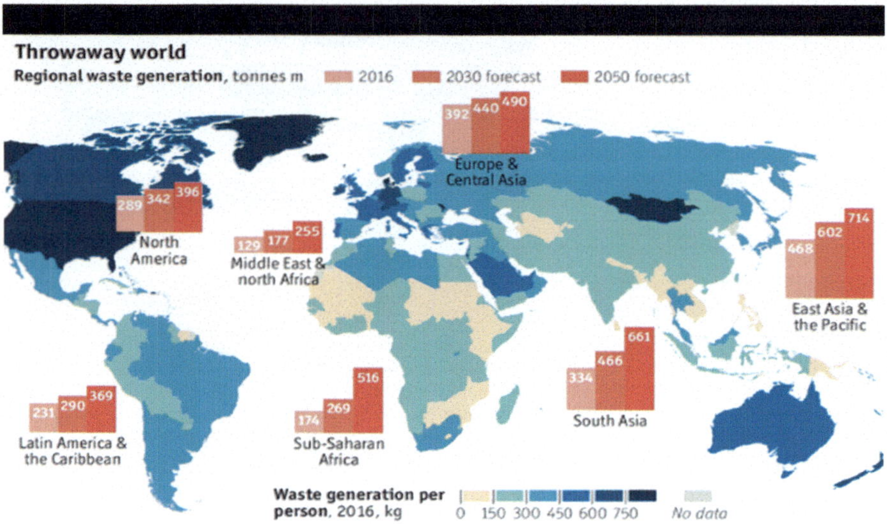


**Fig. 2** Global food waste distribution and forecast

recovery and energy production, which can also help with the ever-increasing global energy demands.

According to the predictions of global food waste as shown in Fig. 2, the volumes of food waste will have increased substantially in 2050 with food waste being projected to be more than twice the current amount. This would mean amongst others that the demand for food worldwide is expected to increase annually and that will put more burden in agricultural sector [15].

The forecasts shown in Fig. 2 suggest that more food waste will be generated in sub-Saharan Africa, soon. Since South Africa has the higher gross domestic product (GDP) per capita as compared to its Sub-Saharan counterparts, this would imply more food waste will be generated in South Africa as there is a strong correlation evident between a country's GDP and waste generation [6].

Regarding future food security, studies have shown that halving the total amount of food calories currently lost and wasted from 24 to 12% by 2050 would save roughly 1,314 trillion kcal of food per year. This will result in 22% of the estimated calories needed to feed the world population in 2050, based on business-as-usual projections for food demand [16].

## 3 Impact of Food Waste on the Environment

The food waste generally poses a myriad of negative impacts on the environment, resources (such as water, fertilizers, pesticides, seeds, energy, labor and capital) and also has economic implications. Of the food waste recovery hierarchy levels, landfilling and incineration are two preferred approaches. Unfortunately, these approaches have adverse environmental impacts due to leaching in landfill sites that will result in a detrimental effect on the aquatic life. With incinerated and disposed of food waste causing the serious emissions of methane gas ($CH_4$), which is 23 times more potent than carbon dioxide ($CO_2$) as a greenhouse gas thereby significantly contributing to climate change [7]. Therefore, practicing landfills as a disposal solution for the treatment of food waste could lead to further global warming.

Further note that food and agricultural organization of the UN's (FAO) in 2011 estimated the food waste to emit 3.6 $GtCO_2eq$ with the exclusion of 0.8 $GtCO_2eq$ of deforestation and managed organic soils associated with the food wastage. Therefore, the total carbon footprint arising from food waste with the inclusive of land use change, is Ca. 4.4 $GtCO_2eq$ per year, equivalent to approximately 7% of global greenhouse gas emissions. Is should be noted that the greenhouse gas emissions of the production phase (including all agricultural inputs, machinery, livestock, soils) and successive phases (such as processing, transportation, preparation of food, waste disposal) were all included in their calculation. A summary showing the contribution of each commodity to carbon footprint and food wastage is depicted in Fig. 3. It is therefore clear that one kg of wheat, or one kg of beef, has different carbon footprints, since their life cycles are different, emitting specific types and varying amount of greenhouse gases [17].

In 2005, agriculture accounted for 10–12% of the total global anthropogenic greenhouse gas emissions, while it accounted for about 60% of $N_2O$ and about 50% of $CH_4$ [18]. The impact of food related greenhouse gases emissions is expected to increase congruently in the near future because of increasing world population and in the consumption of animal products [19, 20].

According to Garnett et al. [21], the levels of greenhouse gas emissions associated with food production are on par with those produced by the transport sector which

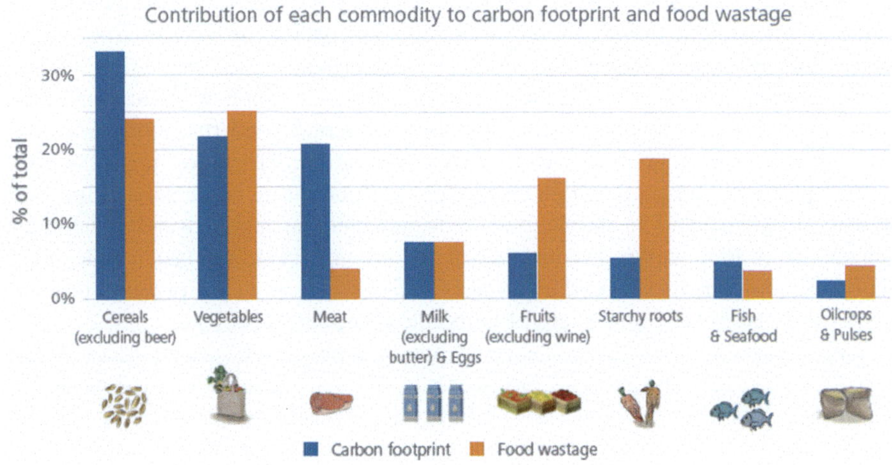


**Fig. 3** Contribution of each commodity to carbon footprint and food wastage [17]

is usually seen as one of the major greenhouse gas contributors. Of the one-fifth of all greenhouse emissions generated in the UK and Australia were linked to food production and consumption. However, the drive to develop efficient food waste management is underway, driven by rising organic waste to produce animal feed and fertilizers, reducing greenhouse emissions and growing energy demand. Based on the projections, the global food waste management market is optimistically estimated to expand with a CAGR of 5.0% from USD 32.64, which is the previous estimation in 2018 [22] even though the incineration and landfill techniques is expected to have a detrimental effect on the market growth.

## 4 Food Waste Characteristics

UN's FAO systematically categorises food waste into eight commodities including cereal, starchy roots, oil crops and pulses, fruit, meat, fish and sea food, milk, eggs and vegetables [23]: Of note is that defining food waste is not always straightforward as distinguishing between edible and nonedible parts of food is subjective [24]. In this chapter though, nonedible part will refer to food waste since in some parts of the world, a food judged edible may be considered nonedible in other parts. Nonedible food waste such as rotten fruits and vegetables, fish and poultry organs, intestine, meat trimmings and other residues, fruits and vegetable peelings, meat, fish, shellfish shells, bones, food fats, sauces, condiments, soup pulp, herbal medicinal pulp, eggshells, cheeses, ice cream, yogurts, tea leaves, teabags, coffee grounds, bread, cakes, biscuits, desserts, jam cereals of all types e.g., rice, noodles, oats, plate scrapings and leftover of cooked food, cooked left overs, and different pet foods [1, 25, 26].

According to Xu et al. [27], food waste is generated from various segments of food supply chain from the food production industry (postharvest residuals such as fruit and vegetable residuals); food processing and packaging (wastes from breweries, slaughterhouses, vegetable peels, etc.); distribution and marketing (supermarket expired food, etc.), and the consumption sector (kitchen waste, meal leftovers, etc.). Despite posing disposal and environmental concerns, food waste has attracted a lot of attention due to its rich composition. Therefore, understanding specific physical and chemical characteristics of food waste is crucial as it plays a pivotal role when deciding among food waste management approach and/or valorization options of food wastes. For instance, food waste mainly consists of a wide range of organic components with carbohydrates (35.5–69%), 3.9–21.9% proteins, oils and fats, organic acids [28], and traces of inorganic compounds [29].

With this being said, the presence of optimum levels of carbohydrates in the form of free sugars and proteins are necessary for fermentation of food waste [2, 30], a high lipid content could be inhibitory to anaerobic digestion [31] but oils can be esterified with alcohol to produce biodiesel, sugars can be esterified for use as surfactants whilst starches can be esterified for use as biodegradable plastics, coatings, and hot-melt adhesives [32]. Also, note that specific properties like pH are fundamental parameters for all biochemical valorization options, and heating value of the food waste is a primary factor in thermochemical processing operations, which is a function of elemental composition [33]. Further note that protein waste from the food industry can come from a variety of sources including, but not limited to, dairy, oilseeds, grains, soybeans, and eggs. Proteins from waste streams can be hydrolysed by proteases into bioactive peptides or commodity chemicals (e.g., polymer precursors) [34]. According to Zan et al. [35] the moisture content of food waste obtained from different sources typically contain varies from 66–80% whilst the organic from variation ranges from 80–94%. In this context, Karmee et al. [1], cited the dependency of industrial production of biofuel from food waste on factors such as availability of food waste, efficiency of hydrolysis process, the amount of lipid and carbohydrate obtained from food waste, and efficiency of fermentation (biofuel) and transesterification (bioethanol) methods.

## 5 Biofuel

There is a potential growth in biofuel owing to awareness regarding renewable energy [2, 36], deadly air pollution [37], reducing the economic dependence on fossil fuels [18, 37–40] as well as decreasing the emissions of air pollutants and greenhouse gases such as nitrogen oxide, sulfur oxides and methane [41]. As compared to their fossil fuels counterpart, biofuels possess a galore of benefits which encompass the abundance of biomass sources, biodegradability, environmentally friendly potential and lower threat to the ecosystem, renewability and contribution to sustainability, beneficial for the economy including extending the opportunities for agriculture, increasing industrial investment, boosting agricultural income, creating rural manufacturing jobs and achieving energy security [36].

Succinctly, biofuels can be described as liquid, solid and gas fuels (such as wood pellets, biogas, ethanol and biodiesel) which are used to deliver renewable energy services in the form of heating and cooling, electricity and transportation [41, 42]. There are various biomass sources that can be converted to biofuel products via thermochemical, physicochemical and biochemical processes [38, 43]. For 1st generation biofuels, sugarcane and corn feedstocks are not able to fulfill the current demand for alcohol. Non-edible lignocellulosic biomass is an alternative bio-source for creating 2nd generation biofuels and algae biomass for 3rd and 4th generation biofuels [44].

It is said that the choice of process technology depends solely on the product desired and the feedstocks [38]. The transesterification and alcoholic fermentation processes are used respectively to produce biodiesel and bioethanol [1, 38].

## 6 Biodiesel

Due to its environmental benefits and the fact that it is made from renewable resources, biodiesel has become more attractive recently. The production of biodiesel using alternative feedstocks to conventional fossil raw materials such as lignocellulosic raw materials was major evolution. Apart from offering a tremendous environmental benefit, biodiesel has another advantage of the good fuel properties i.e., high flash point, good lubricity, etc. [45]. Biodiesel is defined as a mono-alkyl ester of a long fatty acid or fatty acid methyl ester derived from vegetable oils or animal fats and alcohol with or without a catalyst [2, 46] is a renewable and clean-burning diesel replacement for petroleum [2].

Interesting enough though, there are four main routes by which raw vegetable oil and/or animal fat can be converted into biodiesel and these are direct use or blending of oils, micro-emulsion, thermal cracking or pyrolysis and transesterification reaction [2, 46]. Of the aforesaid four routes, the transesterification method is the preferred way to produce biodiesel from diverse feedstock types owing to the quality of the fuel produced [46]. Figures 4 and 5 demonstrate the chemical reaction and process undertaken to manufacture the biodiesel by the transesterification process from waste vegetable oils or animal fats, which involves the reaction of the tri-, di-and monoglycerides with methanol in the presence of a catalyst to produce biodiesel [1, 47]. During this process, the exchange of the organic group can be accelerated by a catalyst. In a general sense, the reaction happens through three consecutive reversible reactions. In the first reaction diglyceride is produced from triglycerides, then monoglyceride is produced from diglyceride in the second reaction, and finally glycerin is produced from monoglycerides in the third reaction. An esters molecule is produced in all three reactions. Therefore, at the end of the reaction, three molecules of alkyl esters and one molecule of glycerin are obtained as shown in Fig. 4. Stoichiometrically 1 mol of fatty acid is required to react with 3 mol of alcohol to produce 3 mol of the methyl ester and 1 mol of glycerol. However, in general a higher alcohol molar ratio is applied to maximize the ester production depending on the oil, type of catalyst, and process parameters [48].

In biodiesel production the transesterification reaction is carried out in a batch reactor between triglyceride and alcohol. In this process, parameters such as mixing intensity, catalyst, temperature, and alcohol and its ratio to oil play a pivotal role as

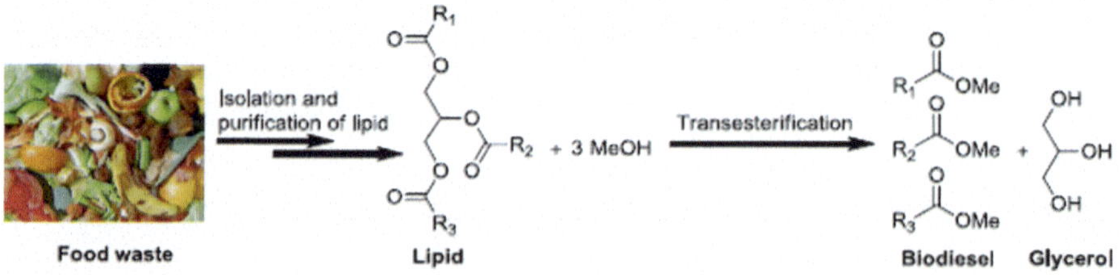


**Fig. 4** Isolation and purification of lipid followed by its transesterification to produce biodiesel [1]

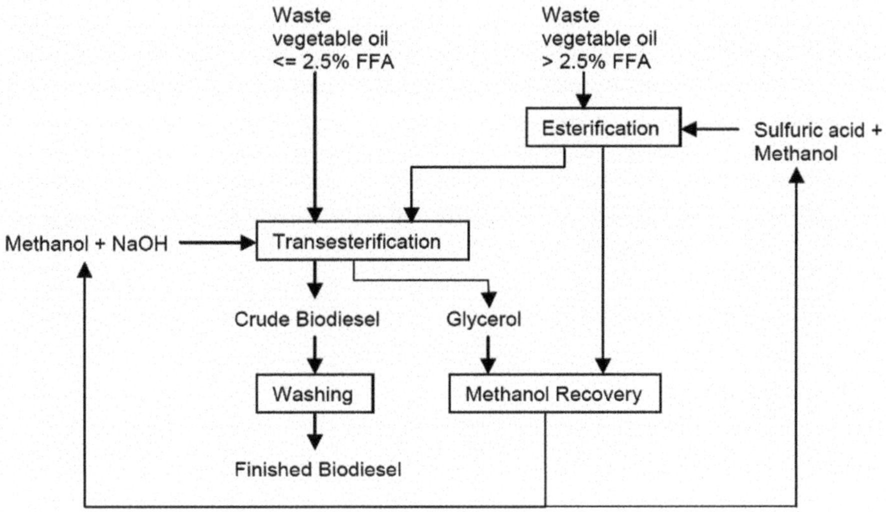


**Fig. 5** Flow diagram of the biodiesel making process [47]

they affect the process yield. [49, 50]. Several studies have reported an increase in the yield by playing around with one of the said parameters. For example, Degfie et al. [51] used the waste cooking palm oil used for food cooking at cafés, restaurants and street fast food in Addis Ababa. A CaO nano-catalyst, with a mean particle size of 29 nm, was synthesized by thermal decomposition method and used as a catalyst for biodiesel production in the transesterification process from waste cooking oil. The optimal biodiesel yield of 96% was achieved at optimized reaction conditions i.e., waste cooking oil to methanol molar ratio of 1:8, 1 wt. % of CaO nano-catalyst, 50 °C reaction temperature and 90 min reaction time. According to these researchers, the produced biodiesel viscosity, specific gravity, water and sediment, total acidity, ash content and sulfur content were tested according to the American fuel standards (ASTM D 6571) and found in good agreement of the standard. Whereas Babaki and coworkers [52] obtained an elevated level of biodiesel from waste cooking oil using a multi-enzyme system developed by covalently immobilizing Lipase (from *Rhizomucor miehei*) and lipase B (from *Candida antarctica*) onto epoxy-functionalized silica. The influences of different factors such as enzyme ratio, the ratio of *t*-butanol to oil, water adsorbent content and reaction duration were studied and optimized by response surface methodology under a central composite design. A high yield of fatty acid methyl esters (91.5%) was recorded after ten hours of reaction time.

One of the highly emphasized advantages of the food waste biodiesel is its ability to be blended with petroleum diesel and used in vehicles engines without need for modification. Biodiesel is often used as a blend, B20 (20 vol% biodiesel and 80 vol% conventional diesel), rather than B100. Seeing the steady supply of waste coffee, biofuel company bio-bean and petrol firm Shell teamed up to produce a biofuel made from waste coffee grounds for some of Transport for London's huge fleet buses. The B20 (a biofuel blend) biofuel contains a 20 per cent bio-component, partly made up of coffee oil, and can be incorporated into the London bus fuel supply chain without the need for modifications to any of the buses. Of note is that with the average Londoner drinking 2.3 cups of coffee a day, producing more than 200,000 tonnes of waste a year. Bio-bean collects some of the waste coffee grounds left behind from high street chains, such as Caffe Nero, and factories and saves them from going to landfill, a waste stream that has the potential to emit 126 million kilogrammes of $CO_2$ [53].

## 7 Bioethanol

Bioethanol is environmental friendly fuels produced by fermentation of carbohydrates produced in sugar or starch crops such as corn, sugarcane, or sweet sorghum [54, 55] including wheat, potatoes and sweet potatoes [56]. Since the major components of food waste are starch, cellulose and proteins materials, it would imply that starch and cellulosic components of the food waste can be hydrolysed to monomeric sugars which can be used in the ethanol production [57]. According to Lopes et al. [58], starch and sugar crops have been dominant in the United States and Brazil's ethanol fuel industry which account for at least 85% of the global market such that corn ethanol conversion both by dry or wet mill processing is widely used in the United States and mostly concentrated in the Midwest area. Due to conducive weather conditions for sugarcane plantation, in Brazil, sugarcane remains the Brazilian primary feedstock in ethanol industry. Despite though, reports suggest that the use of sugar and starch as raw materials for ethanol production had a negative effect on their availability as human food as this raised their prices globally and increased concern about the ethics of biofuel production. This has led to a change of focus on the use of renewable raw material i.e., lignocellulosic biomass and waste [59, 60] as a prime agricultural land normally required for producing foodstuffs not necessarily be for fuel production.

Reports further suggest that organic wastes such as fruit and vegetable have biological and chemical potential for producing bioethanol owing to percentage of indigenous microorganism such as *Candida spp.* and chemical component resulting to their high complex saccharide in a form of lignocellulose. In this context, high source of lignocellulose could be hydrolyzed into D-glucose and D-xylose which could further convert bioethanol by microorganism [61]. Ethanol production generally utilizes the separate hydrolysis and fermentation process, which includes three distinct steps such as liquefaction (or preparation of the fermentable sugar-containing solution), saccharification and fermentation [62]. Enzymatic hydrolysis has also been recommended for the hydrolysis of organic waste during bioethanol production to sugars as this increase the number of sugars that will be available for fermentation [63]. Finally, distillation is done to obtain pure ethanol.

**Fig. 6** Enzyme saccharification of food waste using carbohydrates

During the saccharification process, fermentable sugars such as glucose, fructose, xylose, maltose, amylose, sucrose and arabinose can be produced for subsequent ethanol production. This is achievable through the use of commercial enzyme cocktails used directly or a synergistic mixture of α-amylase, β-amylase, cellulases, xylanases, pullulanase and glucoamylase of various origins used in food waste saccharification [2, 64, 65]. The overall process of fermentative ethanol production is shown in Figs. 4 and 5. Succinctly, Figs. 4 and 5 depict the stages which the food waste undergo during the fermentation process (Fig. 6).

Suffice to say, it is for over a decade now that global ethanol production was exploited via the fermentation process by *Saccharomyces cerevisiae* and remains as prime species. For example, Moon et al. [65] conducted laboratory experiments to evaluate a new potential application of food waste as an alternative substrate for ethanol production. Here food waste was hydrolysed using individual enzymes, their cocktails, batch ethanol fermentation by *Saccharomyces cerevisiae* and the effect of salt, which is inherently included in food waste, on ethanol fermentation. According to these researchers, a higher glucose yields on the pretreated food waste pretreated with an amyloglucosidase, carbohydrase, and a mixture of both enzymes of 0.41, 0.35 and 0.46 g/g respectively. However, a high ethanol yield of 0.23 g/g of dry food waste was obtained after 15 h of fermentation by *Saccharomyces cerevisiae* by using the food waste both hydrolysed by the enzyme mixture and was estimated to be nearly equivalent to the ethanol yields of lignocellulose biomasses. Yan et al. [64] also evaluated the enzymatic hydrolysis of food waste by commercially available enzymes and the subsequent ethanol fermentation of the hydrolysates by *Saccharomyces cerevisiae* H058. In this study, the optimum batch enzymatic conditions were found to be saccharification pH of 4.5, temperature of 55 °C, glucoamylase concentration of 120 u/g, α-amylase concentration of 10 u/g, solid–liquid ratio of 1:0.75 (w/w). Here, fed batch hydrolysis was started with a solid–liquid ratio of 1:1 (w/w), with solid food waste added at time lapse of 2 h to get a final solid–liquid ratio of 1:0.5 (w/w). After 4 h of reaction, the reducing sugar concentration reached 194.43 g/L with an enzymatic digestibility of 93.12%. Further fermentation of the batch and fed batch enzymatic hydrolysates, which contained reducing sugar concentration of 131.41 and 194.43 g/L respectively, was performed using *Saccharomyces cerevisiae* H058 and the yield of 62.93 and 90.72 g/L ethanol was obtained within 48 h. The overall chemistry process for the fermentation process is to convert glucose sugar ($C_6H_{12}O_6$) to ethyl alcohol ($CH_3CH_2OH$) and $CO_2$.

Also note that, Chan and coworkers [66] evaluated the effects of feeding yeast with different glucose concentrations on the ethanol production by batch and fed-batch cultures in a 5 L fermentor. According to these researchers, when a batch culture system with *Saccharomyces cerevisiae* was used for ethanol fermentation with glucose concentrations ranging 10–260 g/L, as a result, 0.2–7.0 g/L biomass and 5.1–115.0 g/L ethanol were obtained even though the substrate inhibition was observed with the initial glucose concentrations greater than 200 g/L in the fermentative media. When a fed-batch culture system (an initial glucose concentration of 180 g/L and total glucose concentration of 260 g/L) was performed, the maximum ethanol concentrations and ethanol yield were significantly higher than those of the batch cultures. The cell biomass, maximum ethanol concentration, and ethanol yields for the fed-batch fermentation cultures were 8.3 g/L, 130.1 g/L and 51% (100% of the theoretical value), respectively. In their studies, Akpan et al. [67], extensively evaluated the conversion of organic waste (old newspapers) and food waste using acid and microbial hydrolysis which yielded 42 and 63% fermentable sugar. Using *Saccharomyces cerevisiae* in the fermentation process, a yield of 95% ethanol was obtained. Unlike other researchers, Prasoulas et al. [68] employed mixture of *Fusarium oxysporum* and *Saccharomyces cerevisiae* cultures to boost fermentation of various agroindustrial residues. Here a higher bioethanol conversion of 20.6 g/L was evident when mixed culture was supplemented with glocuamylase as compared to monoculture of *Fusarium oxysporum* where 16.3 g/L was attained. Interestingly, a further increase in the yield of bioethanol (to 30.3 g/L) was achieved by supplementing the mixed culture with glocuamylase. It is imperative to also state that enzymatic hydrolysis is said to be the costliest step in the production of bioethanol. It significantly increases the total cost of the production of bioethanol and was identified as a barrier in the further deployment of ethanol production [69].

With all being said, owing to its basic properties viz. higher heats of vaporization, higher flame speed, and higher octane number, bioethanol is undoubtedly given attention on a global scale and is being considered as a long term gasoline replacement that would help with the reduction of exhaust emissions. One of the major advantages of bioethanol is that they can be blended with gasoline to form gasohol [70, 71]. Interestingly, Khoung et al. [72] evaluated the effect of –gasoline-bioethanol blends using a fully synthetic oil which was homogenously mixed with five formulated fuels viz. gasoline blend (E0), gasoline–10% ethanol (E10), gasoline–20% ethanol (E20), gasoline–30% ethanol (E30), and gasoline–85% ethanol (E85). These mixtures were then tested in a four-ball wear tester according to the ASTM D4172 standard test. These authors reported that under selected operating conditions, the addition of a gasoline–bioethanol blend decreases the oil viscosity, whereas the acid number increases because bioethanol is more reactive compared to gasoline resulting in enhanced oil degradation and oxidation. Further note that it was also reported that fuel dilution reduces the lubricating efficiency and the wear protection of the engine oil. All fuel-diluted oil samples have higher friction and wear losses, compared to the fresh synthetic oil. E10 has slight effects on the friction and wear behaviors of engine oil. Similar observations were reported by Wibowo et al. [73] where the effect of the bioethanol-gasoline blend on performance SI engine 150 CC was evaluated.

Here the combustion was unstable, torque and power were better in high percentage ethanol blends. In conclusion these authors suggested engine modification for the gasohol for better combustion even though the value performance of the engine can be decreased. It is also reported elsewhere [71] that the most common way to overcome this problem is to blend ethanol with a small fraction of a much more volatile fuel such as gasoline; the most popular blend is E85 which consists of 85% ethanol and 15% gasoline by volume. Although E85 has been extensively used worldwide, engine manufacturers guarantee problem-free operation without any modification only to catalyst equipped cars fueled with gasoline containing no more than 5% ethanol. However, modern catalyst-equipped cars are probably able to run without any material problem with up to 20% ethanol which seems to be the upper limit for cold climates.

It worth noting that most cars on the road today in the U.S. can run on blends of up to 10% ethanol, and the use of 10% ethanol gasoline is mandated in some U.S. states and cities. Whereas, In Brazil blending of ethanol with gasoline has been mandatory since 1976, and since 2007 the legal blend is around 25% ethanol and 75% gasoline (E25). Most interestingly, by 2010 Brazil had a fleet of more than 10 million flexible fuel vehicles regularly using neat ethanol fuel already [74].

## 8 Biogas

Reports suggest that food waste disposed of landfill emits a large volume of methane of about 40–70% and carbon dioxide [7, 75]. Using the anaerobic digestion process food waste can be converted into biogas. Anaerobic digestion process is a biological process that in the absence oxygen, the microorganism degrades the complex organic matter to simpler components under anaerobic conditions to produce biogas and fertilizer [76, 77]. Anaerobic digestion is said to be a superior alternative to composting as it is a net-energy producer [76]. Biogas is a mixture rich in methane that can be used as combustible gas, electricity and heat generation, connection to the natural gas grid or as a biofuel in vehicles as shown in Fig. 7 [76–81]. Figure 8 shows the schematic diagram of anaerobic digestion process. That is to say, directing food waste through anaerobic digestion, to produce biogas in a controlled conditions, has environmental benefits and promotes circular economy [75].

Under anaerobic conditions, biogas is formed by degradation of organic matter and the quality of the resultant biogas depends heavily on the feedstock and production technology [77]. The anaerobic process involves three mayor phases i.e., hydrolysis, fermentation and methanogenesis [38]. In contrast, Nsair et al. [76] and [77] who stated that the biodegradation of the complex matter undergoes four main steps viz. hydrolysis, acidogenesis, acetogenesis and methanogenesis. According to Lee et al. [38], in hydrolysis step the complex biomolecules in the biomass are broken down into simple biomolecules and fermentation using the simple biomolecules to construct alcohols, acetic acid, fatty acids that are volatile, and $H_2$ and $CO_2$

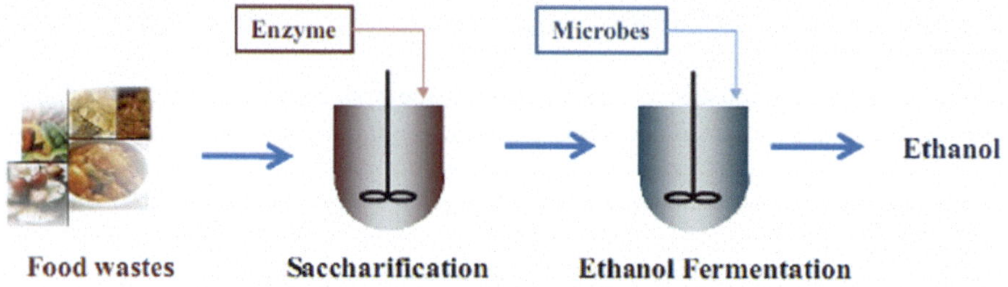


**Fig. 7** Separate enzyme saccharification and ethanol fermentation

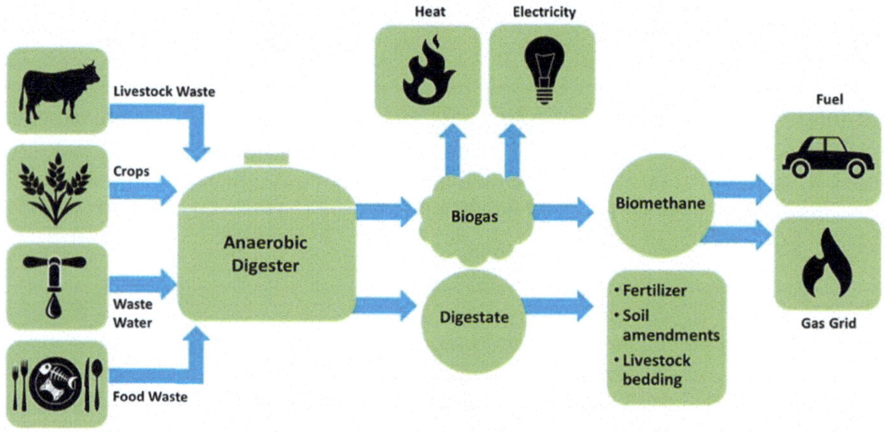


**Fig. 8** Anaerobic digestion process (Graphic by Sara Tanigawa, EESI) [81]

gas mixture whilst methanogens metabolized this gas mixture producing biogas comprising $CH_4$ (60–70%) and $CO_2$ (30–40%).

During the fermentation of organic materials takes place within large tanks called anaerobic digesters, there are requisites for metabolic product of the participating methanogens and acidogenic bacteria that need to be harnessed. The pro qua non for their production conditions include a lack of oxygen, a pH value from 6.5 to 7.5 and a constant temperature of 35–45 °C (mesophilic) or 45–55 °C (thermophilic). Generally, the digestion period or retention period is between 10 and 30 days depending upon the type of digestion employed. The anaerobic digestion systems of today operate largely within the mesophilic temperature range [76]. Similarly, Zamandeh et al. [82] evaluated the performance and microbial structure of a digestion using food waste-only and a co-digestion process (mixture of cow manure and food waste) at mesophilic (37 °C) and thermophilic (55 °C) temperatures. According to the aforesaid researchers, the mesophilic digested with food waste gave the highest methane (480 mL/g VS) whilst the mesophilic co-digestion of the mixture of food waste and manure yielded 26% more than the sum of the individual digestions of manure and food waste. In this study, the dominant bacteria in mesophilic digester were *Chloroflexi* (54%), while *Firmicutes* were dominant in mesophilic mixed co-digestion of food waste and manure (60%).

Interesting enough though is that biogas is gaining popularity across the world owing to its several benefits, various types of biogas plants have been developed in different parts of the world. For example, in Mexico a new biogas plant erected on the Bordo Poniente (a former Mexico City's largest landfill site) was expected to be commission in 2020. The project was because of concerted effort between Mexico development banks Bancomext and Nafinsa, and national infrastructure fund Fonadin. Report suggests the plant will provide an efficient alternative for treating the waste from this city of 10 million people and also generate 965,000 MWh of electricity per year, which will be used directly by the Mexico City Subway Metro. In 2017, the biogas plant with a capacity to generate 290 GW hours of electricity was built in Warsaw in Duplin County. The facility in question was designed to process more than 750,000 tons of organic waste and power 32,000 homes annually. Further note that Carbon Cycle Energy (C2e) of Colorado (a company constructing America's largest utility-scale biogas plant agreed to supply the plant's total output of bio-methane to Duke Energy and to an undisclosed 'Fortune 500 company'. Similar biogas facilities were built in countries such as Demark, Sweden, UK, and Spain to process agricultural byproduct and organic waste to produce biogas even though they differ in terms of size, capacity and the output [83]. Of note is that, in Western Cape South Africa, there is a plan which is in the finalisation stage to build the biogas plant in the near future. The identified places for building the biogas facilities will include amongst others a piggery (for a 500 kW electrical energy application), while the plant at the second piggery will be for a combination of heat (400 kW) and electrical power (75 kW). The plant for the dairy farm will be a 250 kW electric power application [84].

## 9 Current Developments (Covid Case)

Apart from being a threat to worldwide health since its outbreak in early December 2019 [85], SARS-CoV2 (that causes COVID-19) pandemic has generated an unprecedented impact in most countries of the world [86], and continues to wreak havoc globally [87]. To curb the spread of COVID-19 as there is no specific vaccines or treatment for this global pandemic [88], most countries were ordering lockdowns, restricting movement, implementing massive COVID-19 screening tests and establishing public policies of social distancing [85–88] to protect their population [88]. Though the ongoing COVID-19 pandemic has resulted in many unintended, but not unwelcome, environmental benefits as skies clear and animals venture into newly empty space [89], the pandemic has disrupted the whole food supply chain [85–89], from farm to fork [90] thereby shifting food access, food security and food losses due to changes in food flow and distribution patterns [91].

However, COVID-19 pandemic has stressed the relevance of performing a deep review regarding the fit-for-purpose of current food production consumption systems [85] as the exceptional lockdown measures somewhat exposed the loopholes on the entire food supply chain. With this in mind, for many citizens the lockdown measures have overtly or covertly prompted an accelerated learning process of food purchase management perhaps due to socio-economical context of the COVID-19 crisis such as food availability, restricted movements and/or loss of income than by a pro-environmental concern [88]. The perils of this new plague have encouraged the need for collective efforts and actions to stem the problem of food waste [87]. Interestingly, the COVID-19 related interruptions and/or detrimental impact within the food supply chain around the world present researchers and governments with the opportunity to rethink food provision and build resilient, sustainable and democratic agri-food systems that would provide food democracy [92].

## 10 Conclusion

Inadequate waste management remains a huge threat to social, economy and environment as many countries still use incineration and landfill to dispose of food waste. Because for decades landfills have allowed us to ignore our waste. Unfortunately, food waste discarded in landfills should be avoided as they cause greenhouse gas emissions and groundwater contamination. Individuals can reduce their food waste contribution through simple adjustments to consumption habits. However, this requires establishment social awareness of food waste and its impacts thereof. Social awareness campaigns are gradual processes that cannot achieve its intended results instantaneously especially when consumers are expected to change their habits. While the consumers can adopt sustainable and conscientious practices, food waste system is large and complex and food waste are generated throughout the food supply chain. This would imply that food waste can only be reduced but it cannot entirely be eliminated via social awareness campaigns and changed consumption habits. Note that some citizens learned the prominence of food management during the COVID-19 lockdown period, however, the amount of food waste generated during the period had increased by 12%. This shows that food waste can be used as a sustainable feedstock for volarisation of food waste into biofuels such as biodiesel, bioethanol and biogas. This will help reduce the toxic greenhouse emissions emitted by food waste in landfills and also replacement of fossil fuel-based fuels by renewable fuels.

In conclusion, thorough studies need to be done to validate the sustainability of food waste as feedstock of valorization process in Sub-Saharan Africa as the predictions indicate more food waste will be generated in this region soon. This can be achieved, inter alia, by introducing the food waste bins in university canteens, hospitals, restaurants, abattoirs, piggeries, and chicken farms across the countries to enable proper quantification of food waste generated weekly. This will also help in terms one to discern whether food waste varies by monthly, seasons, semester and so forth. The information gathered can play an important role to investors who wish to build one of voralisation and biomass plants. The project itself needs the consented efforts from both research and development, government and private sector most especially in South Africa where the economy is heavily reliant on fossil fuel-based energy.

## References


1. Karmee, S. K., & Lin, C .S .K. (2014). Volarisation of food waste to biofuel: Current trends and technological challenges. *Sustainable Chemical Processes*, (22), 1–4.
2. Shuang, L., & Yang, X. (2016). Biofuel production from food wastes. In *Handbook of biofuels production* 2*e*.
3. Retrieved August 17, 2020, from https://publications.parliament.uk/pa/cm201617/cmselect/cmenvfru/429/429.pdf.
4. Närvänen, E., Mesiranta, N., Mattila, M., & Heikkinen, A. (2020). Introduction: A framework for managing food waste. *Food Waste Management*, 1–24.
5. Retrieved August 17, 2020, from https://www.nrdc.org/sites/default/files/wasted-food-IP.pdf.
6. Oelofse, S. H. H., & Nahman, A. (2013). Estimating the magnitude of food waste generated in South Africa. *Waste Management and Research, 31*(1), 80–86.
7. Pham, T. P. T., Kaushik, R. K., Parshetti, G. K., & Mahmood, R. (2015). Food waste-to-energy conversion technologies: Current status and future directions. *Waste Management, 38*, 399–408.
8. Papargyropoulou, E., Wright, N., Lozano, R., Steinberger, J., Padfield, R., & Ujang, Z. (2016). Conceptual framework for the study of food waste generation and prevention in the hospitality sector. *Waste Management*, (49), 326–336.
9. House of Commons, Environment, Food and Rural Affairs Committee. (25 April 2017). *Food waste in England*. Eight Report of Session 2016–2017.
10. Towards the future we want: End hunger and make the transition to sustainable agricultural and food systems. Food and agriculture organization of the United Nations (FAO, 2012).
11. Retrieved September 14, 2020, from https://www.epa.gov/sites/production/files/2016-01/documents/msw_task11-2_foodwastemanagementscopingstudy_508_fnl_2.pdf.
12. EPA. (2016). U.S. Environmental Protection Agency. Sustainable management of food. Retrieved June 17, 2013, from https://www.epa.gov/sustainable-management-food.
13. Bušić, A., Kundas, S., Morzak, G., Belskaya, H., Marđetko, N., Šantek, M. I., Komes, D., Novak, S., & Šantek, B. (2018). Recent trends in biofuel and biogas production. *Food and Technology, 56*(2), 152–173.
14. Sheahan, M., & Barret, C. B. (2017). Review: Food loss and waste in Sub-Saharan Africa. *Food Policy, 70*, 1–12.
15. Ching-Hsu, H., Shih-Min, L., & Nai-Yun, H. (2020). Understanding global food surplus and food waste economic and environmental sustainability. *Sustainability, 2892*(12), 1–18.
16. Kennard, N. (2019). *Chapter: Food waste management.*
17. Retrieved September 15, 2020, from http://www.fao.org/3/a-bb144e.pdf.
18. Smith, P., Martino, D., Cai, Z., Gwary, D., Janzen, H., Kumar, P., et al. (2007). Policy and technological constraints to implementation of greenhouse gas mitigation options in agriculture. *Agriculture, Ecosystems and Environment*, 118, 6–28.
19. Geeraert, F. (2013). Sustainability and dietary change: The implications of Swedish food consumption patterns 1960–2006. *International Journal of Consumer Studies, 37*, 121–129.
20. Abeliotis, K., Costarelli, V., & Anegnostopoulos, K. (2016). The effect of different types of diet on greenhouse gas emissions in Greece. *International Journal on Food System Dynamics, 7*(1), 36–49.
21. Garnett, T. (2009). Livestock-related greenhouse gas emissions: Impacts and options for policy makers. *Environmental Science and Policy, 12*(4), 491–503.

22. Retrieved September 17, 2020, from https://www.grandviewresearch.com/industry-analysis/food-waste-management-market.
23. Food and Agriculture Organization of the United Nations. (2014). *Food wastage footprint*.
24. Stafford, W. H. L., Lotter, G. A., von Maltitz, G. P., & Brent, A. C. (2019). Biofuels technology development in Southern Africa. *Development Southern Africa, 36*(2), 155–174.
25. Retrieved September 22, 2020, from https://www.nrdc.org/sites/default/files/food-waste-city-level-report.pdf.
26. Retrieved September 22, 2020, from https://www.epd.gov.hk/epd/sites/default/files/epd/english/environmentinhk/waste/prob_solutions/files/Code%20of%20Practice%20for%20FW%20collector_final_14%20July%202017.pdf.
27. Xu, F., Li, Y., Ge, X., Yang, L., & Li, Y. (2018). Anaerobic digestion of food waste−challenges and opportunities. *Bioresource and Technology*, 247, 1047−1058.
28. Paritosh, K., Kushwaha, S. K., Yadav, M., Pareek, N., Chawade, A., & Vivekanand, V. (2017). Food waste to energy: An overview of sustainable approaches for food waste management and nutrient recycling. *Biomed Research International*, 1–20.
29. Kiran, E. U., Trzcinski, A. P., Ng, W. J., & Liu, Y. (2014). Bioconversion of food waste to energy: A review. *Fuel, 134*, 389–399.
30. Hegde, S., Lodge, J. S. J. S., & Trabold, T. A. T. A. (2018). Characteristics of food processing wastes and their use in sustainable alcohol production. *Renewable and Sustainable Energy reviews, 81*, 510–523.
31. Chen, Y., Cheng, J. J., & Creamer, K. S. (2008). Inhibition of anaerobic digestion process: A review. *Bioresource Technology, 99*(10), 4044–4064.
32. Andler, S. M., & Goddard, J. M. (2018). Transformation food waste: How immobilized enzymes can Varolise waste streams into revenue streams. *Science of Food*, 1–11.
33. Mahmood, R., Parshetti, G. K., & Balasubramanian, R. (2016). Energy, exergy and technoeconomic analyses of hydrothermal oxidation of food waste to produce hydro-char and bio-oil. *Energy*, (102), 187–198.
34. Zan, F., Dai, J., Hong, Y., Wong, M., Jiang, F., & Chen, G. (2017). The characteristics of household food waste in Hong Kong and their implications for sewage quality and energy recovery. *Waste Management*, 1–11.
35. Gog, A., Roman, M., Toşa, M., Paizs, C., & Irimie, F. D. (2012). Biodiesel production using enzymatic transesterification–current state and perspectives. *Renewable Energy*, (39), 10–16.
36. Kamani, M. H., Eş, I., Lorenzo, J. M., Remize, F., Roselló-Soto, E., Barba, F. J., et al. (2019). Advance in plant materials, food by product, and algae conversion biofuels use of environmentally friendly technologies. *Green Chemistry*, 12.
37. Ouadi, M., Bashir, M. A., Speranza, L. G., Jahangiri, H., & Hurnung, A. (2019). Food and market–a pathway to sustainable fuels and waste volarization. *Energy & Fuels, 33*, 9843–9850.
38. Lee, S. Y., Sankaran, R., Chew, K. W., Tan, C. H., Krishnamoorthy, R., Chu, D.-T., & Show, P.-L. (2019). Waste to bioenergy: A review on the recent conversion technologies. *BMC Energy, 1*(4), 1–22.
39. Hirani, A. H., Javed, N., Asif, M., Basu, S. K., & Kumar, A. (2018). A review on first- and second-generation biofuel. *Biofuels: Greenhouse Gas Mitigation and Global Warming*, 141–154.
40. Kim, E. J., Kim, S., Choi, H. G., & Han, S. J. (2020). Co-production of biodiesel and bioethanol using psychrophilic microalgae *Chlomydomonas* sp. KNM0029C isolated from Arctic sea ice. *Biotechnology for Biofuels, 13*(20), 1–13.
41. Cushion, E., Whiteman, A., & Dieterle, G. (2010). *Bioenergy development: Issues and impacts for poverty and natural resource management*. The International Bank for Reconstruction and Development/The World Bank.
42. Ghosh, P. R., Fawcett, D., Sharma, S. B., & Poinern, G. E. J. (2016). Progress towards sustainable utilisation and management of food wastes in the global economy. *International Journal of Food Science*, 1–22.
43. Kouzu, M., & Hidaka, J.-S. (2012). Transesterification of vegetable oil into biodiesel catalyzed by CaO: A review. *Fuel, 93*, 1–12.

44. Sheldon, R. A. (2018). The road to biorenewables: Carbohydrades to commodity chemicals. *ACS Sustainable Chemistry and Engineering, 6*, 4464–4480.
45. Mishra, V. K., & Goswami, R. (2017). A review of production, properties and advantages of biodiesel. *Biofuels, 9*(2), 273–289.
46. Gebremariam, S. N., & Marchetti, J. M. (2017). Biodiesel production technologies: Review. *Aims Energy, 5*(3), 425–4577.
47. Retrieved September 23, 2020, from https://www.istc.illinois.edu/research/energy/biofuels_biodiesel.
48. Hobuss, C. B., da Silva, F. A., Dos Santos, M. A., de Pereira, C. M., Schulz, G. A., & Bianchini, D. (2020). Synthesis characterization of monocylglycerols through glycerolysis of ethyl esters derived from linseed oil by green processes. *Journal of RSC Advances*, 4.
49. Gónzalez, J. P., Gutiérrez, P. E. Á., Medina, M. A., Zapata, B. Y., Guerrero, G. V. R., & Valdés, L. G. V. (2020). Effect on biodiesel production caused by feed oil changes in a continuous stirred-tank reactor. *Applied Sciences, 10*(992), 1–24.
50. Alsoudy, A., Thomsen, M. H., & Janajreh, I. (2012). Influence on process parameters in transesterification of vegetable and waste oil-a review. *IJRRAS, 10*(1), 1–15.
51. Degfie, T. A., Mamo, T. T., & Mekonnen, Y. S. (2019). Optimized biodiesel production from waste cooking oil (WCO) using calcium oxide (CaO) nano-catalyst. *Scientific Reports, 9*(18982), 1–8.
52. Babaki, M., Yousefi, M., Habibi, Z., & Mohammadi, M. (2017). Process optimization for biodiesel production from waste cooking oil using multi-enzyme systems through response surface methodology. *Renewable Energy, 105*, 465–472.
53. Retrieved September 25, 2020, from https://resource.co/article/biofuel-made-waste-coffee-grounds-power-london-buses-today-12254.
54. Byadgi, S. A., & Kalbulgi, P. B. (2016). Production of bioethanol from waste newspaper. *Procedia Environment, 35*, 555–562.
55. Gavahian, M., Munekata, P. E., Eş, I., Lorenzo, J. M., Khaneghah, A. M., & Barba, F. J. (2019). Emerging techniques in bioethanol production from distillation to waste varolization. *Journal of Green Chemistry*, 6.
56. Sheikh, R. A., Al-Bar, O. A., & Soliman, M. Y. A. (2016). Biochemical studies of the production of biofuel (bioethanol) from potatoes peels wastes by saccharomyces cerevisiae: Effects of fermentation periods and nitrogen source concentration. *Biotechnology and Biotechnological Equipment, 30*(3), 497–505.
57. Thongdumyu, P., Intrasungkha, N., O-Thong, S. (2014). Optimisation of ethanol production from food waste hydrolysis by co-culture of Zymomonas Mobilis and Candida Shehatae under non-sterile condition. *African Journal of BioTechnology, 13*(7), 866–873.
58. Lopes, M. L., de Paulillo, S. C. L., Godoy, A., Cherubin, R. A., Lorenzi, M. S., Giometti, F. H. C., et al. (2016). Ethanol production in Brazil: A bridge between science and industry. *Brazilian Journal of Microbiology, 47S*, 64–76.
59. Motsakas, L., & Christakopoulos, P. (2015). Ethanol production from enzymatically treated dried food waste using enzymes produce On-site. *Sustainability, 7*(2), 1446–1458.
60. Bušić, A., Marđetko, N., Kundas, S., Morzak, G., Belskaya, H., Ivančić Šantek, M., et al. (2018). Bioethanol production from renewable raw materials and its separation and purification: A review. *Food Technology and Biotechnology, 56*(3), 289–311.
61. Utama, G. L., Sidabutar, F. E. E., Felina, H., Wira, D. W., & Balia, R. L. (2019). The utilization of fruit and vegetable wastes for bioethanol production with inoculation of indigenous yeasts consortium. *Bulgarian Journal of Agricultural Science, 25*(2), 264–270.
62. Dey, P., Lhakpa, W., & Singh, J. (2015). Simultaneous saccharification and fermentation (SSF), an efficient process for bioethanol production: An Overview. *Biosciences Biotechnology Research Asia, 12*(1), 87–100.
63. Nair, R. B., Lennartsson, P. R., & Taherzadeh, M. J. (2016). Bio ethanol production from agricultural and municipal wastes. *Current Developments in Biotechnology and Bioengineering; Solid Waste Management*, 157–188.

64. Yan, S., Yao, J., Yao, L., Zhi, Z., Chen, X., & Wu, J. (2012). Fed batch enzymatic saccharification of food waste improves the sugar concentration in the hydrolysates and eventually the ethanol fermentation by saccharomyces cerevisiae H058. *Brazilian Archives of Biology and Technology, 55*(2), 183–192.
65. Moon, H. C., Song, I. S., Kim, J. C., Shirai, Y., Lee, D. H., Kim, J. K., et al. (2009). Enzymatic hydrolysis of food waste and ethanol fermentation. *International Journal of Energy Research, 33*, 164–172.
66. Chang, Y. -H., Chen, C. -Y., Hsu, C. -L., Chang, T. -C., & Jang, H. -D. (2018). Enhancement of the efficiency of bioethanol production by Saccharomyces cerevisiae via gradually batch-wise and fed-batch increasing the glucose concentration. *Fermentation, 4*(2), 45.
67. Prasoulas, G., Gentikis, A., Konti, A., Kalantzi, S., Kekos, D., & Mamma, D. (2020). Bioethanol production from food waste applying the multienzyme system produced On-site by *Fusarium oxysporum* F3 and mixed microbial cultures. *Fermentation, 6*(2), 39.
68. Akpan, U. G., Alhakim, A. A., & Ijah, U. J. (2008). Production of ethanol from organic and food waste. *Leonardo Electronic Journal of Practices and Technologies, 13*, 1–11.
69. Padella, M., O'Connell, A., & Prussi, M. (2019). What is still limiting the deployment of cellulosic ethanol? Analysis of the current status of the sector. *Applied Science, 9*(4523), 1–14.
70. Retrieved September 29, 2020, from https://www.car-engineer.com/adapting-an-engine-to-ethanol-fuel-flex-fuel/.
71. Retrieved September 29, 2020, from https://ec.europa.eu/jrc/en/publication/eur-scientific-and-technical-research-reports/effect-fuel-ethanol-content-exhaust-emissions-flexible-fuel-vehicle.
72. Khuong, L. S., Masjuki, H. H., Zulkifli, N. W. M., Mohamad, E. N., Kalam, M. A., Alabdulkarem, A. et al. (2017). Effect of gasoline–bioethanol blends on the properties and lubrication characteristics of commercial engine oil. *RSC Advances*, 25.
73. Wibowo, C. S., Sugiarto, B., Zikra, A., Budi, A., Mulya, T., & Muchar, M. (2019). The effect of bioethanol varying gasoline blends on performance and emission of SI engine 150 cc. In *AIP Conference Proceedings of the 10th International Meeting of Advances in Thermofluids* (*IMAT2018*) (Vol. 2062).
74. Yesmin, M. N., Azad, M. A. K., Kamruzzaman, M., & Uddin, M. N. (2020). Bioethanol corn, pumpkin and carrot of Bangladesh as renewable source using yeast Saccharomyces Cerevisiae. *Acta Chemica Malaysia (ACMY), 4*(2), 1–10.
75. Morales-Polo, C., Cledera-Castro, M. D. M., & Moratilla Soria, B. Y. (2019). Biogas production from vegetable and fruit markets waste-compositional and batch characterizations. Sustainability, 11, 6790.
76. Retrieved September 29, 2020, from https://www.clarke-energy.com/applications/biodegradblewaste/.
77. Nsair, A., Cinar, S. O., Alassali, A., Qdais, H. A., & Kuchta, K. (2020). Operational parameters of biogas plants: A review and evaluation study. *Energies, 13*(3761), 1–27.
78. Dung, T. N. B., Sen, B., Chen, C.-C., Kumar, G., & Lin, C.-Y. (2014). Food waste to bioenergy via anaerobic processes. *Energy Procedia, 61*, 307–312.
79. Rosén, T., & Ödlund, L. (2019). System perspective on biogas use for transport and electricity production. *Energies, 12*(4159), 1–12.
80. Dahlgren, S. (2020). Biogas based fuels as renewable energy in the transport sector: an overview of the potential of using CBG and LBG and other vehicle fuels produced from biogas. *Biofuels*, 1–14.
81. Retrieved September 29, 2020, from https://www.eesi.org/papers/view/fact-sheet-biogasconverting-waste-to-energy.
82. Zamanzadeh, M., Hagen, L. H., Svensson, K., Unjordet, R., & Horn, S. J. (2017). Biogas production from food waste via co-digestion and digestion effects on performance and microbial ecology. *Scientific Reports, 7*(17664), 1–12.
83. Retrieved September 30, 2020, from https://www.nsenergybusiness.com/news/newslargest-biogas-plants-061017-5943061/.

84. Retrieved September 30, 2020, from https://www.engineeringnews.co.za/article/food-processing-industry-ideal-for-biogas-production-2019-07-30/rep_id:4136.
85. Aldaco, R., Hoehn, D., Laso, J., Margallo, M., Ruiz-Salmón, J., Cristobal, J. et al. (2020). Food waste management during the COVID-19 outbreak: A holistic climate, economic and nutritional approach. *Science of the Total Environment, 742*, 140524.
86. Wu, F., Zhao, S., Yu, B., Chen, Y. M., Wang, W., Song, Z. G., & Yuan, M. L. (2020). A new coronavirus associated with human respiratory disease in China. *Nature, 579*(7798), 265–269.
87. Mitigating risks to food systems during COVID-19: Reducing food loss and waste. Food and Agriculture Organisation of the United Nations 11 May 2020.
88. Jribi, S., Ben Ismail, H., Doggui, D., & Debbabi, H. (2020). COVID-19 virus outbreak lockdown: What impacts on household food wastage? *Environment, Development and Sustainability*, 1–17.
89. Retrieved October 01, 2020, from http://environmath.org/2020/06/25/the-many-faces-of-food-waste/.
90. Retrieved September 30, 2020, from https://farmdocdaily.illinois.edu/2020/09/food-waste-and-covid-19-impacts-along-the-supply-chain.html.
91. Shurson, G. C. (2020). "What a waste"–can we improve sustainability of food animal production system by recycling food waste streams into animal feed in an era of health, climate, and economic crises? *Sustainability, 12*(7071), 1–34.
92. Petetin, L. (2020). The COVID-19 crisis: An opportunity to integrate food democracy into post-pandemic food systems. *European Journal of Risk Regulation, 11*, 326–336.

The manufacturer's authorised representative in the EU is Springer
Nature Customer Service Centre GmbH, Europaplatz 3, 69115 Heidelberg,
Germany. If you have any concerns regarding our products, please
contact ProductSafety@springernature.com

Printed and bound by CPI Group (UK) Ltd, Croydon, CR0 4YY
01/10/2026
02225945-0010